JOURNAL OF CHROMATOGRAPHY LIBRARY - volume 21

environmental problem solving using gas and liquid chromatography

JOURNAL OF CHROMATOGRAPHY LIBRARY

Volume 1 Chromatography of Antibiotics, by G.H. Wagman and M.J. Weinstein

Volume 2 Extraction Chromatography, edited by T. Braun and G. Ghersini

Volume 3 Liquid Column Chromatography. A Survey of Modern Techniques and Applications, edited by Z. Deyl, K. Macek and J. Janák

Volume 4 Detectors in Gas Chromatography, by J. Ševčík

Volume 5 Instrumental Liquid Chromatography. A Practical Manual on High-Performance Liquid Chromatographic Methods, by N.A. Parris

Volume 6 Isotachophoresis. Theory, Instrumentation and Applications, by F.M. Everaerts, J.L. Beckers and Th.P.E.M. Verheggen

Volume 7 Chemical Derivatization in Liquid Chromatography, by J.F. Lawrence and R.W. Frei

Volume 8 Chromatography of Steroids, by E. Heftmann

Volume 9 HPTLC — High Performance Thin-Layer Chromatography, edited by A. Zlatkis and R.E. Kaiser

Volume 10 Gas Chromatography of Polymers, by V.G. Berezkin, V.R. Alishoyev and I.B. Nemirovskaya

Volume 11 Liquid Chromatography Detectors, by R.P.W. Scott

Volume 12 Affinity Chromatography, by J. Turková

Volume 13 Instrumentation for High-Performance Liquid Chromatography, edited by J.F.K. Huber

Volume 14 Radiochromatography. The Chromatography and Electrophoresis of Radiolabelled Compounds, by T.R. Roberts

Volume 15 Antibiotics. Isolation, Separation and Purification, edited by M.J. Weinstein and G.H. Wagman

Volume 16 Porous Silica. Its Properties and Use as Support in Column Liquid Chromatography, by K.K. Unger

Volume 17 75 Years of Chromatography — A Historical Dialogue, edited by L.S. Ettre and A. Zlatkis

Volume 18 Electrophoresis. A Survey of Techniques and Applications. Part A: Techniques, edited by Z. Deyl

Volume 19 Chemical Derivatization in Gas Chromatography, by J. Drozd

Volume 20 Electron Capture. Theory and Practice in Chromatography, edited by A. Zlatkis and C.F. Poole

Volume 21 Environmental Problem Solving using Gas and Liquid Chromatography, by R.L. Grob and M.A. Kaiser

JOURNAL OF CHROMATOGRAPHY LIBRARY - volume 21

environmental problem solving using gas and liquid chromatography

Robert L. Grob

Professor of Analytical Chemistry, Villanova University, Villanova, PA 19085

Mary A. Kaiser

Supervisor, Separations Group, Central Research and Development Department, E.I. du Pont de Nemours & Co., Experimental Station, Wilmington, DE 19898

ELSEVIER SCIENTIFIC PUBLISHING COMPANY
Amsterdam – Oxford – New York 1982

ELSEVIER SCIENCE PUBLISHERS B.V.
Sara Burgerhartstraat 25
P.O. Box 211, 1000 AE Amsterdam, The Netherlands

Distributors for the United States and Canada:

ELSEVIER SCIENCE PUBLISHING COMPANY INC.
52, Vanderbilt Avenue
New York, NY 10017

First edition 1982
Second impression 1985

ISBN 0-444-42065-7 (Vol. 21)
ISBN 0-444-41616-1 (Series)

Printed in The Netherlands

CONTENTS

PREFACE

"A question to ponder is this: would this large 'home' we call the environment survive if we leave it completely in the hands of Mother Nature? If the answer is yes, does this allow us to be completely unconcerned? If the answer is no, then how far should we extend our studies and safeguards? The authors feel that this is not an easy yes or no question, but that the answer lies in finding an equilibrium somewhere between the two extremes. Once this point of equilibrium is found, it is simply a matter of adjusting the variables from time to time so that we don't shift to the 'reactant' or 'product' sides and become incompatible with the system." (R.L. Grob, Environmental Studies of the Atmosphere by Gas Chromatography, in Contemporary Topics in Analytical and Clinical Chemistry, D.M. Hercules, G.M. Hieftje, L.R. Snyder, and M.A. Evenson (Eds.), Plenum Press, New York, 1977).

Environmental analyses can be justified for many reasons: (1) consideration of the population because of residential atmospheric pollution, (2) monitoring of technical advances to control pollution, (3) assessing plant damage and animal injuries because of environmental damage to our agricultural systems, (4) identifying substances which explain events or activities which have taken place in the environment and (5) preserving the work areas in industries by detecting harmful components which can be injurious to personnel.

Air and water pollutants may be gases, liquids or solids which can change the natural composition of our environment. Air pollutants can be classified as emissions if they escape into the outside air after being discharged from their pollution site or as emissions if they stay near the site of generation.

In the United States smog results from catalysis by solar radiation of olefins and nitrogen oxides, whereas in Europe most of the smog results from the sulfur dioxide released by flue gases.

We can place the effects of air pollution in better perspective if we consider that a normal breath is approximately 0.5 liter and a deep breath can be from 1.5 to 2.0 liters. Thus, on the average we inhale about 6-10 m^3 of air/day (i.e. 4.2-7.0 ℓ/min). The nose, mouth, and to some extent the trachea and bronchia will retain particles which are 5-40 μm in diameter but particles 0.5-5 μm in diameter usually enter the lungs and are deposited. Particles less than 5 μm (fine dusts) are usually exhaled.

Accuracy of an environmental analysis means its correctness. Sensitivity refers to the smallest amount of a contaminant that can be detected with certainty. It is the task of the analytical chemist to select and sometimes modify the method which will be applied so that specific requirements of accuracy and sensitivity are met and relevant information is obtained. The number of different methods available are wide in scope and the appropriate procedure must be chosen to satisfy the needs of the problem. The environmental problem may concern property damage and less elaborate sampling procedures and less accurate methods may provide the information. However, when the environmental problem involves a health hazard it is not wise to allow economic considerations to preclude reliability and comprehensiveness. The number of different methods and instruments available make it necessary for the analytical chemist to choose the correct system so that he can produce relevant data.

The authors have chosen the course that sampling per se is a technique which occurs outside the laboratory (i.e., obtaining a representative portion of the bulk population), whereas any manipulation of the sample inside the laboratory falls into the realm of sample treatment. As a consequence of this decision, headspace (vapor equilibration) sampling fits more appropriately into sample treatment, i.e., in the general category of extraction treatment of laboratory samples. We realize that a number of readers and workers in the field of environmental analysis will not agree with this categorization; however, it does have its merits.

(1) It removes the overlap or grey area of what is actually sampling in the true sense and sample treatment.

(2) It clearly establishes the fact that what one does with the "representative" sample in the laboratory is really "preparing" it for analysis.

(3) When the analytical chemist performs "headspace sampling," "vapor equilibration," and/or "purge and trap" techniques he is only treating a sample that must be representative of the bulk population of sample available to him.

All written works certainly have shortcomings. Nothing is perfect in the world of reality, and anything that endeavors to be absolutely ideal could never be completed. Our completed work, its style or overall format will not fill the requirements of all persons. One writes what he considers the best manuscript to serve the majority of the intended readership. Constructive comments are solicited, but we recognize that agreement among all is impossible.

The completion of a written work is not possible without the assistance and comments of others who are qualified to review the manuscript. The authors wish to thank their respective spouses, Cecil (MAK) and Marge (RLG) for being understanding during the time this was being written. We also wish to thank Dr. Gerald R. Umbreit for reading and commenting on a number of chapters and Norman Henry III and Joseph Viskocil for their comments on chapter 7. We appreciate the assistance of Mary Ann Quarry, Mary Ellen McNally, Proespichaya Kanatharana, and Sithichai Leepipatpiboon in proofreading the final copy. To Dr. Cecil Dybowski go our special thanks for his valuable editorial assistance with this manuscript. We acknowledge

Linda Grob for the illustrations, Betty B. Wolfe, Sharon M. Dize and Dorothy Lauder for setting the manuscript into print and being patient with all the changes and last minute additions, and finally the support of the DuPont Company in this connection.

CHAPTER I

SCOPE OF THE PROBLEM

1.1 HISTORY AND BACKGROUND

Environmental problems have burdened mankind since the first human roamed the earth. As technology progressed and man's appetite for a carefree life expanded, the environmental problems became more complex and difficult to alleviate. Lodge (ref. 1) suggested that the first air-pollution problem may have occurred when a prehistoric man brought a burning branch into his cave. Certainly such an event could have been viewed as discomforting to the cave's other occupants who could have filed the first environmental impact statement on the desirability of bringing smoky branches into enclosed areas. Three categories of pollution are our biggest concern: water pollution, air pollution and solid-waste disposal. Water pollution has been present longer than the other two and it is a more widespread problem. Modern technology is finding ways to dispose of or use solid waste and to reduce the effluents from industrial stacks and auto-emissions; however, it is very expensive to clean up polluted water and/or find a market for its pollutants.

It is difficult to imagine what may have been the first water-pollution problem. Perhaps it was associated with the natural pollution of lake or river by an overgrowth of plants or contamination by human waste. It is interesting to note that the book of Deuteronomy (Chapter 23: Verses 12-14) contains one of the earliest references to construction and placement of sanitary facilities for human waste to avoid both air and water contamination.

For years man used a small percentage of the natural resources (trees, minerals, ore deposits, animals and fish) available to him. The discovery and use of fire, disposal of his waste products and work processes had little or no effect on the rain, the content of rivers and streams, land erosion, and life cycles of animals and fish. These natural resources were able to replenish themselves because they were in greater abundance than man. The increase in population and technology has caused a large shift in this situation. We have disposed of hundreds of millions of tons of garbage and trash into a few dumps and landfills, released hundreds of billions of gallons of waste-containing waters into our streams and rivers and sent hundreds of millions of tons of pollutants into our atmosphere. (The atmosphere nearest to our Earth consists of the troposphere and the stratosphere. The troposphere is that portion of the atmosphere from Earth's surface to an altitude of seven miles (11 km); the stratosphere begins at seven miles and extends to 17 miles (27 km) above the Earth's surface).

Fuller (ref. 2) has presented a unique correlation between energy consumption and the number of persons on the Earth. "The Egyptian pharoahs who built the great pyramids commanded the labor (energy) of 100,000 slaves. The glories of Athens in its Golden Age were supported by the labor of 125,000 slaves. Life in the United States today is smoothed by the energy equivalent to the labor of 40 billion slaves. To keep us warm, manufacture our goods, grow and process our foods, transport these necessities and luxuries to our homes, and entertain us in our leisure hours, each of us commands the energy equivalent to the work of 200 full-time slaves."

It has been estimated (ref. 3) that if our present mode of life continues, nonrenewable natural resources will become exhausted and the death rate will rise sharply.

With the onset of high fossil fuel use in the 13th and 14th centuries, air pollution became a serious problem in urban areas. The first bans on the burning of coal came during the rein of Elizabeth I when Parliament banned coal burning in London city limits while Parliament was in session. As a matter of fact, in most countries, smoke and soot problems spurred the first pollution legislation. The rate of consumption of fossil fuels continues to increase. A reasonable estimate is that today fossil fuel is being used a million times faster than it can be formed.

Often environmental legislation followed serious air-pollution incidents such as the inversions (regions of the atmosphere where the temperature increases with altitude) in the Meuse River Valley in Western Europe in 1930, Donora, PA in 1948 (four-day inversion period killed 20 persons and caused more than 7000 persons to be ill), and in London in 1952 (four-day inversion period killed 4000 persons) (ref. 4). The air-pollution episode in Donora was one of the most carefully studied cases in modern history. It took many years of investigation and research to conclude that zinc ammonium sulfate probably caused the illnesses among 43 of the persons in the Donora area. The most probable source of the pollution was the local steel and zinc plants.

Many factors can enter into air-pollution crises. It is not within the scope of this book to detail all of these except to say that meteorological, topographical and some light intensity factors play important roles. Earth rotation distorts natural air circulation because mountains and canyons deflect air flow. Clouds and haze absorb more heat than clean air, where air becomes warmer over barren ground than air over ground with heavy vegetation. Also air over water is cooler than ground air. All these variables affect the flow of air; air movement is limited. Due to the slow movement of weather fronts and mountain barriers, little cloud cover is present to reduce the ultraviolet rays of the sun. This results in a huge chemical reaction basin being formed in which photochemical reactions can occur. Thus, in order to minimize the potential for these reactions, one must control the type and quantity of effluents in the basin.

Acid rains cause a great deal of damage to our buildings, soil, plant and animal life. Belgium and The Netherlands have the most acidic rain in the world. These probably result from the fact that they are centrally located in the European industrial complex.

A number of terms frequently appear in the air-pollution literature, and these should be familiar to the scientist working with environmental problems. A number of pollutants released into the atmosphere will interact with other pollutants, and it is these reactions which determine the final concentration of many pollutants. An example of this phenomenon is the classical formation of peroxyacylnitrate which occurs when hydrocarbons react with oxygen atoms from the decomposition of nitrogen dioxide to form free radicals. These free radicals react with oxygen and nitrogen dioxide molecules to form the peroxyacylnitrates.

The maximum concentration of a pollutant allowed in the air without causing health ill-effects is known as tolerance level. The residence time of a pollutant is the time necessary for the pollutant to be removed by natural purification processes. The greenhouse effect is the heating of the earth's atmosphere because the atmosphere transmits the visible and ultraviolet light from the sun and absorbs the infrared light generated at the earth's surface. Large concentrations of carbon dioxide in the atmosphere will enhance this effect. The result of this effect is that the earth's temperature has increased because the average carbon dioxide concentration has increased during the past century.

Air pollutants are basically any volatile materials generated at the Earth's surface (oxides of carbon, sulfur, nitrogen; hydrocarbons; gaseous metal compounds, etc.) and particulate matter (a general term which includes small solid particles, as well as liquid droplets).

One of our most fragile natural sources is the water supply. Through the course of human development, the availability of water determined the location for settlement and programs of civilization. If a settlement did not maintain a clean water supply, the inhabitants were forced to move on to a new location. As the population of the earth grew, fewer places remained to which groups were able to move, and it became imperative to maintain good, clean sources of water. The growth of technology has made it possible to utilize water for manufacturing of goods and also for removal of contaminants which cause human health problems. While most manufacturers acted responsibly in the utilization of this natural resource, a few did not. The result of these irresponsible acts has been detrimental not only to man, but to all plants and animals.

The increase in the extent to which water is used in our advanced technological society suggests great potential for accelerating water crises. In the U.S.A. alone, it is estimated that 1×10^{12} ℓ (300×10^9 U.S. gallons) of water are withdrawn yearly from lakes, rivers and streams (ref. 5). Over 90% of this water is used for industrial purposes and much of it is used and discharged as waste water. This water may be changed significantly and possibly create health problems. Characteristics of water which are cited in water-quality criteria are dissolved oxygen; solids (sludge, floating solids, scum, particulates); color; turbidity; bacteria, protozoa, and viruses; taste; odor; salts (including ammonia, cations, anions); pH; temperature; organic compounds; pesticides; herbicides; and radioactive materials. Aquatic species have adapted to the natural organic materials in the water, and even some of the recent man-made organic materials; however, man has increased his production of organic wastes which enter our water system as industrial by-products or waste, e.g. fuels, plastics, detergents, solvents, dyes, inks, paint, drugs, pesticides, and herbicides. Unless there is a spill, most of these pollutants are present in small amounts (as far as economical reclaiming is concerned) and cause toxic problems to man, animals, and fish. Many of these products are not biodegradable (able to be broken down by bacteria), and are passed on from fish to animal and man. Oil films on water prevent the oxygen and carbon dioxide exchange between air and water and in turn kill many waterfowl and fish. Water pollution and its many resulting problems could be reduced by three main requirements:

(1) All municipal sewage be given primary, secondary and tertiary purification treatments.

(2) Industrial waste be reasonably clear of such trace pollutants before being passed into our waterways.

(3) Industrial plants consider the recycling of water.

Subsequent chapters in this book will outline the role of chromatography for the analysis and identification of some of our environmental challenges.

1.2 METHODOLOGICAL QUESTIONS

The effects of our efforts to protect the environment are directly proportional to our capabilities of measuring the level of changes in our environment. Popular awareness of environmental problems is needed; however, the responsibility for establishing an accurate environmental baseline, for setting and implementing standards and for monitoring environmental change must rest with the analytical chemist. This responsibility is awesome and must carry with it a deep moral commitment for the careful implementation of all aspects of environmental/analytical chemistry. The analytical chemist must consider every aspect of samples in the light of the problem at hand from selection of the sampling site to the final data interpretation.

In order to measure any species of interest, often it is necessary and desirable to separate it from a complex matrix. The most common separations tools for the analytical chemist are gas and/or liquid chromatography. It is with these techniques that this book is concerned as they relate to environmental problem solving. For a comprehensive survey of gas and liquid chromatograhy, the reader should refer to the books by Snyder and Kirkland (ref. 6), Grob (ref. 7) and Touchstone and Rogers (ref. 8).

The selection of a separation method is dictated by the instrument availability, the number of samples, the required accuracy and precision, sensitivity of the method, the amount of sample preparation needed and the cost, not necessarily in that order. If only one type of instrument is available, the only problem may be to demonstrate that separation can be carried out to the desired specification. Information on the accuracy, precision and sensitivity of the method must be checked by experiment, although it can often be predicted from the literature data. The sensitivity often relies on the capability of the detectors (see Chapters 5 and 6). The sample preparation time depends on the nature of the problem and the depth of information needed. After the initial purchase cost of the instrumentation, labor is the highest budget item. It is obvious that much consideration must be given to these criteria, and the procedures outlined well in advance of the actual laboratory analysis. Most standard methods for chromatographic analyses of the environment involve gas chromatography. In general gas chromatography is a less expensive technique to use than liquid chromatography, is more familiar to most chemists, and has the advantage of being capable of separating 20% of the known organic compounds without prior modification of the sample. Gas chromatographic detectors are generally more selective than liquid chromatographic detectors. Liquid chromatography has the advantage that it can separate mixtures at lower temperatures, vary both the mobile and stationary phase for increased selectivity, and can use a variety of unique column packings. The two techniques are complementary and can be automated, although gas chromatography may present fewer automation problems (especially as far as equilibration, liquid chromatographic gradient elution, and mobile phase disposal are concerned). The next seven chapters lay the foundation for the complete attack on environmental problems using the two popular analytical techniques, gas and liquid chromatography. Tables 1.1 and 1.2 provide the baseline information which one needs to compare components of "clean air and water." Chapter 8 contains extensive information on the regulations by various governmental agencies.

TABLE 1.1

Composition of dry air[a] (ref. 9)

Element	Concentration (%,v/v)	Element	Concentration(ppm) (ppm,v/v)
N_2	78.09	H_2	0.5
O_2	20.95	Kr	1
CO_2	0.03	He	5.2
Ar	0.93	Ne	18
		CH_4	2.2

[a]Water is usually present from 0.1 to 3% in the atmosphere.

TABLE 1.2

Typical cationic concentration ratios (relative to potassium) in sea and fresh water (ref. 10)

Cation	Concentration(ppm) Sea water	Fresh water
Sodium	50	2
Magnesium	5	1
Potassium	1	1
Calcium	1	5

REFERENCES

1 J. P. Lodge, Jr. in J. N. Pitts, Jr., R. L. Metcalf and D. Grosjean (Eds.), Advances in Environmental Science and Technology, Wiley, New York, 1980.
2 E. C. Fuller, Chemistry and Man's Environment, Houghton Mifflin Co., Boston, MA, 1974, p.2.
3 D. H. Meadows, D. L. Meadows, J. Randers, and W. W. Behrens, The Limits of Growth, Universe Books, New York, 1972.
4 R. G. Bond, C. P. Straub, and R. Prober (Eds.), CRC Handbook of Environmental Control, Vol. 1: Air Pollution, CRC Press, Cleveland, OH, 1972, pp. 149-154.
5 Anonymous, Industrial Wastewater, Office of Research and Development, United States Environmental Protection Agency, Cincinnati, OH, 1980, p. 1.
6 L. R. Snyder and J. J. Kirkland, Introduction to Modern Liquid Chromatography, Wiley, New York, 2nd ed., 1979.
7 R. L. Grob (Ed.), Modern Practice of Gas Chromatography, Wiley, New York, 1977.
8 J. C. Touchstone and D. Rogers (Eds.), Thin Layer Chromatography, Quantitative Environmental and Clinical Applications, Wiley, New York, 1980.
9 R. G. Bond, C. P. Straub, and R. Prober (Eds.), CRC Handbook of Environmental Control, Vol. 1: Air Pollution, CRC Press, Cleveland, OH, 1972, p. 3.
10 M. Waldrop, Chemical and Engineering News, American Chemical Society, Washington, DC, 1980, p. 31.

CHAPTER 2

CRITERIA FOR THE SAMPLING PROCESS

2.1 REQUIREMENTS

2.1.1 General discussion

There is a time-honored saying that the analysis of any sample can only be as good as the sample is representative. What this tells us very concisely is that it matters little how much technique we have accomplished or how expensive and elaborate are our laboratory instruments, the information obtained from the analysis will be meaningless if the sample does not represent the system from which it was obtained. It follows then that a corollary regarding sampling and analysis is that there is little or no difference between a biologically-derived material and an industrially-derived material. The only difference from the analytical viewpoint is the number and amounts of the various components making up these materials. Regarding environmental samples, one may cite the overly simplified statement of Umbreit (ref. 1) to bring across this point: "Anything detected in a water sample which is not water must be considered a pollutant or contaminant. The same statement, of course, is true of air." Fortunately, we all know the problem is not quite that simple.

Before sampling and/or analyzing an environmental sample the analytical chemist must perform some preliminary non-experimental work. Answers to some or all of the following should be known.

(a) What chemical compounds are known or of concern in the sample?
(b) At what concentration levels do they exist?
(c) What are the major components of the sample?
(d) How much sample will be available?
(e) Where and under what conditions will the sample be taken?
(f) Can destructive or non-destructive analyses be used?
(g) What conclusions and/or decisions are to be based on the analysis?

There are a number of review articles which may be good starting points for background information on environmental analysis. A review article for potentially hazardous environmental substances determined by GC/MS was published by Oswald et al. (ref. 2), another for the analysis of organophosphorus insecticides and metabolites by Burchfield and Storrs (ref. 3) and a general review of pesticide residues by Thornburg (ref. 4). A state-of-the-art presentation appeared in the May, June and July 1975 issues of Journal of Chromatographic Science. Saltzman and Cuddeback (ref. 5) and Fishman and Erdemann (ref. 6) have published reviews on air pollution and organics in water analysis, respectively. Weil (ref. 7) has discussed various methods for determination of trace substances in water and effluents. Rosen (ref. 8) has discussed a general outline for the identification of organic pollutants in water. The initial separations were based upon group solubility and adsorption chromatography.

Having decided on a chromatographic technique, a knowledge of the sensitivity limits to be achieved with the detector system is assumed. The size of samples and the degree of concentration required is defined. Sampling procedures must be set, i.e., the number of samples and the method of obtaining the samples must be made. Concentrating the sought component, isolation of it from a variety of other constituents of the sample, partial isolation or concentration of the sought component or chemical conversion to a derivative of different chromatographic behavior are now all possible solutions to the analysis problem.

Finally, standards must be prepared to establish the true minimum level of detection and to calibrate the response of the system. The final measurement step includes choice of chromatographic column, operating temperature, mobile phase composition and detector type.

The analytical chemist should have a thorough knowledge of the most commonly used analytical techniques, operation of the instrumentation at his disposal, and the chemical and physical behavior of gaseous systems, airborne particulates, and aqueous solutions. Finally, he should assure himself that his sampling procedures and number of samples will provide the best results.

The amount (size) of sample needed from a system is inversely related to the sensitivity of the analytical technique used. Time required for the sampling is governed by the sampling system; the system may be capable of a sampling rate of ml/min or m^3/min. Efficiency of the collection system is the controlling factor - an adequate and representative sample must be provided. Sampling time will provide an average concentration level for the interval covered. A number of factors will control the minimum sampling time. These factors include the expected concentration range of the sought component, the rate at which the sampling is to be performed, the minimum detectable quantity of the sought component by the analytical procedure used, and the interpretation or use of the final data. A collection efficiency of at least 75% is necessary for most all environmental analyses (ref. 9). To maintain this level of efficiency it is sometimes necessary to use several pieces of apparatus and/or techniques. If this proves to be necessary, then one must ensure that the collection units do not interfere with each other. When gas-sampling an air sample, it is advisable to remove particulate matter first. This may be done by using a filter ahead of the gas-sampling unit. Caution is advised because of adsorption of the gaseous components on the filter and/or particulates. One can determine the efficiency or effectiveness of the sampling system by use of synthetic environmental samples under controlled laboratory conditions and a wide range of standard concentrations. Inefficient sample collection usually begins to appear as lower and lower concentration standards are used.

One may check the efficiency of various impingers or wash bottles (see Chapter 3) by connecting several of them in series and passing a gaseous sample of known contaminant concentration through the system. Each bottle is subsequently analyzed and if the contaminant uptake is greater than 90% in the first bottle, it may be considered efficient. Efficiency of air sampling approaches 100% if the sample is collected by passage into an evacuated container. However, in this type of sampling, one must consider wall effects of the container.

A final comment about air sampling - air analysis is of no use if the sample is collected while it is raining or snowing. The presence of the liquid phase tends to obscure accurate data for an air sample. Precipitation (rain or snow) is a major source of nutrients. Concentrations of ionic species in the air usually follow a pattern according to sampling area location and the season (ref. 10). The rainwater has distinctive qualities and therefore has an effect on regional, national and continental water supplies. Many of the chemical substances found in streams can be attributed to man's environmental pollution and/or control.

2.1.2 Criteria for sampling

Sampling from either air or water systems is not a simple and straightforward procedure. The analytical chemist must be certain that his samples are directly related to the contaminant concentrations and that his data will substantiate the reasons for collecting samples initially. The fact to keep in mind is that environmental contaminants exhibit a variable concentration and thus are described by normal statistics (ref. 11). An adequate sampling procedure should answer a number of questions:

(a) When does one sample?
(b) Where does one sample?
(c) For what time period does one sample?
(d) How often are the samples taken?

Again, statistics can aid us in our decisions (refs. 12-18) regarding the number of samples, where to sample, etc. However, in many environmental studies it has been the financial limitations of the project rather than experimental design from first principles which dictated the answers to the sampling problem.

In the analysis scheme of environmental samples, it is the sampling step over which the chemist has some control. Thus, the overall scheme may be summed up by the following statements (ref. 19).

(1) Statistical principles will govern the adequacy of the sampling program.

(2) Probability will replace certainty in the results.

(3) Statistical significance will determine the soundness of the conclusions.

2.1.3 Sample types and sampling problem

The analysis of environmental samples presents the analytical chemist a diversified and challenging array of sample types and sampling problems. If the systems to be analyzed are homogeneous, then little or no problems would be present in the sampling step. Under these conditions, we would encounter liquids in liquids (complete miscibility), gases in gases (uniform mixing), gases in liquids (complete solubility), and solids in liquids (complete solubility). The homogeneous system of a solid in a solid would be the last type of sample in this category. This type of sample would be found most commonly in particulate matter samples.

The sample types which may cause sampling problems would be the following:

(A) Aerosols are systems of colloidal particles (10^{-7} to 5 x 10^{-5} cm in diameter which will not diffuse through animal or vegetable membranes) dispersed in a gas, e.g., dusts, soot, particles, fumes, smoke and fogs.
(B) Gels are systems very analogous to sols (see F below) formed by the coagulation of a colloidal liquid.
(C) Foams are a dispersion of a gas in a liquid.
(D) Emulsions are colloidal suspensions (a system consisting of small particles kept dispersed by the molecular motion in the surrounding medium) of liquids in a liquid. The liquid phases are immiscible.
(E) Mists are a suspension of a liquid in a gas, i.e., a cloud-like aggregation of minute globules of water in the atmosphere.
(F) Sols are a colloidal suspension of a solid in a liquid.

The problem-type samples are a cause for caution on the part of the analytical chemist as they all involve colloidal particles. Colloidal particles carry an electrical charge in many cases and have a very large surface area. These two properties tend to cause adsorption of other components on the surface and may result in non-representative sampling and misleading data.

The techniques of sampling are discussed in detail in Chapter 3; however, it is well to point out that these techniques may be modified because of the system being sampled. If the component of most interest varies with the time of day, Time-Proportional Sampling is performed. Taking samples over a prearranged time pattern furnishes more information than one sample taken for a long period of time. In the latter case one obtains an average level for the concentration of the component of interest, whereas in the former case, a more realistic pattern of the fluctuation in concentration results. A similar occurrence can present itself when sampling water from a flowing stream. Because the flowrate may vary with time, it is better to take samples based upon a predetermined volume or to sample at various times during the day rather than on a time basis. This technique is called Flow-Proportional Sampling. This method gives superior information since the sample taken over a period of time can furnish only average concentration levels of the component of interest.

In Section 2.1.1 it is stated that air analysis is of no use if the samples are collected during a period of rain or snow. One could obtain useful information about the effect of rain on air pollution by sampling the rain water. The composition of the rain water is to be considered when one is interested in the composition of soil and irrigation waters (ref. 20). This could be done by use of a rain gauge. If the rain "cleared" the air then one would expect large concentrations of many air pollutants in the rain sample. Rain water will "wash" out and concentrate many organic (including pesticides) and inorganic pollutants of the air (refs. 21-25). The closer to the ocean shores the rain water sample is taken the higher is the salt content.

Rain gauges may be classified as a manual type because the volume reading must be checked periodically, or automatic type. In the automatic recording type, when the compartment is full of rain water, a switch is tripped and an empty compartment is positioned for further collection. A record is kept for the number of times the switch is tripped. Knowing the calibrated volume of the gauge compartment, one has the total volume of rain collected.

A laboratory has been set up at the University of Illinois, Urbana-Champaign, Illinois, to analyze rain-water samples from fifty nationwide sampling stations in the United States. The laboratory is part of the United States National Atmospheric Deposition Program (ref. 26). The field samples may be checked for acidity and conductivity at the sampling site. All other analyses are performed in the laboratory at the University of Illinois.

Humidity can also affect air quality. Low humidity results in increased suspended particulate concentrations and high humidity (fog conditions) blocks solar heating and therefore prolongs the life of inversion layers (ref. 27). Low temperature and fog usually cause increased morbidity and mortality (ref. 28).

Another important sample type is that obtained by Stack Sampling (refs. 29 and 30). This type of sampling is performed either to monitor the efficiency of an industrial collection system or where a suspected or known pollutant has been found in surrounding air. Justification for either of these situations could be:

(1) to determine whether a company is in compliance with local or regional regulations;

(2) to enable a company to select appropriate control equipment for its manufacturing sites;

(3) to identify the presence of a persistent contaminant in the surrounding area;

(4) to determine if present control equipment needs to be changed or improved;

(5) to monitor emissions when a manufacturing process has been changed;

(6) for legal reasons, e.g., to settle a court suit or to answer a governmental regulatory agency citation.

2.2 PROBLEMS OF SAMPLING AND POST-SAMPLES

This section discusses sampling in general, with comments on problems and general precautions. Specific discussion of various techniques will be reviewed in Chapter 3.

The majority of environmental samples are either liquids or gases. It is with these two sample types that we are mainly concerned. Sampling techniques used for surface waters, groundwaters, irrigation and drainage waters are essentially the same. A similar statement may be made of atmospheric air samples, laboratory air samples, stack samples, and general in-house samples. Many of the problems associated with sampling may be classified as those occurring during the "post-sampling" period.

Environmental problems would be very straightforward if all samples were stable for long periods of time. Stability of samples should be checked before analyses are performed. Storage stability can be affected by the storage container, the sample in the container, environmental conditions under which a sample is obtained and stored, concentration of the contaminants of interest, adsorption effects, and presence of other chemical entities in the sample.

Depending upon the source and type of sample, it may be necessary to refrigerate, analyze immediately, use completely filled water sample containers (i.e. leave no headspace), or add a bacterial inhibitor. Volatile materials and dissolved gases will escape from water samples if allowed to be stored for long periods before analysis. In general, filling of the water sample container (no headspace) and refrigeration will minimize these post-sampling changes.

Other post-sampling problems may arise if chemical or physical changes occur. In this instance, having as much advance or additional information regarding the sample source minimizes these problems. The presence of bacteria in a water sample, for example, may cause such changes as reduction of urea to gaseous nitrogen, hydrolysis of urea to ammonia, oxidation of ammonia to nitrite and nitrate, and formation of volatile fatty acids (ref. 20).

Light can accelerate many chemical reactions. Thus, storage in the dark is essential if the samples cannot be analyzed immediately. This is especially true of environmental gas samples where photochemical reactions could result if the samples were stored in direct sunlight.

Another problem with environmental samples which prevents the preservation of integrity of the sample is the loss occurring during separation and/or concentration steps. Due to the low concentration of many contaminants in air and water samples, the analytical chemist must employ some separation and/or concentration technique prior to the measuring step. Disregard for the possible makeup of the sample can furnish sample analysis data which is meaningless. Steam distillation is a handy technique for removing and concentrating volatiles from water; whereas, extraction is a good technique for accomplishing the same end result for neutral or non-volatile components. The reader is referred to a number of references regarding sampling and care of water samples (refs. 31-49).

2.3 REGULATIONS

Specific information regarding the various worldwide regulatory agencies is discussed in Chapter 8. Our concern at this point is more a general statement concerning the regulations on sampling as a whole.

Each of the associations and agencies of the individual countries has requirements which are pertinent to their specific problems. The underlying reasons for all techniques of sampling, types of samples and their subsequent analyses are about the same.

The role of the analytical chemist in all of these projects is very important. He is necessary not only for the analysis per se, but the analytical chemist is vital for the success of resolving environmental problems. If he is to be effective, he must be an integral part of the discussion of the problem initially, the design of the sampling program, the gathering and coding of the samples, the labeling and care of the samples at the completion of the study, and last but not least, the interpretation of the data. He must be capable of discussing and explaining the details and results of such studies to persons not well versed in this area, e.g., other chemists, biologists, physicists, engineers, physicians, the man on the street, judges, jurors, and politicians. Rightfully, the analytical chemist must become completely involved in environmental problems and not follow the narrow path of accepting samples, analyzing these samples, and then passing the data to someone else.

Having established that an environmental problem exists, the next move is to lay out the project so that sufficient information becomes available to arrive at the correct conclusions.

There are a number of organizations which have the responsibility of either setting or controlling regulations pertaining to environmental guidelines. The administrator of the United States Environmental Protection Agency (EPA), for example, must be sure the agency puts forth criteria that reflect the latest scientific knowledge useful in indicating the effects on public health or welfare expected from pollutants in ambient air and industrial waste water effluents. The criteria with which they have been concerned mainly (for air pollution) are particulate matter, sulfur oxides, carbon monoxide, nitrogen oxides, hydrocarbons, polychlorinated biphenyls (PCBs) and other halogenated organics, and photochemical oxidants. For water pollution they have been concerned primarily with pesticides, herbicides, polychlorinated biphenyls and other halogenated organics, toxic trace metals and various industrial wastes. All of these pollutants and methods for their determination are constantly being reviewed.

The Committee on the Challenges of Modern Society (CCMS) established by the North Atlantic Treaty Organization (NATO) set up several air pollution studies. Procedures published by this commission differ from criteria set by the U.S.A (ref. 50).

The National Research Council/National Academy of Sciences (NRC/NAS) published a number of documents reviewing information about many specific substances (ref. 51). These were followed by documents outlining a philosophy for short-term air quality limits (ref. 52).

The World Health Organization (WHO) established a committee on Air Quality Criteria and Guides and published a report entitled Air Quality Criteria and Guides for Urban Pollutants (ref. 53). The present WHO Environmental Health Criteria Programme (ref. 54) was a result of the initial study.

The United Nations Conference on the Human Environment charged WHO with the task of establishing primary standards for pollutants that are common to air, water, and food, and developing procedures for setting derived working limits for common air and water contaminants. "Derived working levels" are also known as environmental or ambient air quality standards, maximum permissible limits, maximum allowable concentrations or product standards.

Air and water pollution abatement are different for every country and within a country from one location or time period to another. If the major health problems arise from an inadequate clean water supply or other sanitation problems, less emphasis can be given to air pollution problems. Conversely, a highly industrialized country may suffer greater air pollution exposure problems than water pollution problems provided waste dumping in streams were controlled.

2.4 STANDARDS

2.4.1 Discussion

Verifications of analytical procedures presuppose the use of standard samples. The standard samples are used to check the correct functioning of analytical instrumentation and procedures. Thus the analytical chemist should be aware of sources of such samples, the cooperative programs which govern these verification samples, and the types of samples and supplies at his disposal.

Commercial and governmental agencies are the main sources of such samples. A number of publications abstract the environmental literature (ref. 55). The EPA publishes Air Pollution Abstracts monthly (ref. 56) and a number of books concerned with source sampling and analyses (ref. 57), monitoring air pollutants (ref. 58), evaluation and detection of aerosols (ref. 59) and monitoring instrumentation and sampling descriptions (refs. 60 and 61) are also available.

Air and water quality standards are legal limits placed on the concentration of water and air pollutants for a given period of time or a given volume. These criteria are decided by consideration of economic, social, technical and political information.

The availability of air and water standards does not preclude the removal of interferences. A number of common analytical techniques are used for the removal of interfering substances. These should be called upon to assist in many analyses. Some of the more routinely used approaches to removal of interferences are:

(1) change in temperature of the sample system;
(2) ashing of residues obtained either from separation of groups of compounds or specific substances;
(3) use of ion exchange to remove ionic interferences;
(4) distillation to remove low boiling components;
(5) complexing agents to mask an inorganic interference;
(6) change in pH of sample system;
(7) use of reaction kinetics to alter amount of interfering substances.

Interferences can occur during various steps in the process of obtaining the desired analytical information. The more common sources would be (a) at the sampling sites themselves, (b) sample collection system could be the problem, (c) reaction which may take place during the sample storage interim, and (d) poor laboratory technique during the analysis. These interferences may not be solely chemical, but may be physical in nature, e.g., temperature effects, light effects (photo-decomposition or photo-oxidation), time interval between sampling step and analysis step, or unclean equipment.

Control of sampling parameters and analysis procedures will still produce indeterminate errors in the data. These may be handled in a statistical manner. The scope of this treatise does not permit the coverage of this topic. The reader is referred to a number of instructive references (refs. 12-18).

2.4.2 Availability and source of samples and supplies

Table 2.1 is a listing of various suppliers of environmental standards for gas and liquid chromatography. Also included are those suppliers which furnish column supplies and reference samples. This is not an all-inclusive listing, but is a list of sources known and/or used by the authors. This listing is not to be interpreted as an approval or recommendation by the authors.

TABLE 2.1

Suppliers of environmental samples, standards, and supplies

Supplier	Product type(s)*
Airco Industrial Gases 575 Mountain Avenue, Murray Hill, NJ 07974, USA.	CRM, RM
Air Products and Chemicals, Inc. Specialty Gas Department P.O. Box 538, Allentown, PA 18105, USA.	CRM, RM, Gas standards, Std. gas samples
Aldrich Chemical Co., Inc. 940 W. St. Paul Avenue, Milwaukee, WI 53233, USA.	GC and LC standards
Aldrich-Europe, Division of Janssen Pharmaceuticals Turnhoutsebaan-30, B2340, Beerse, Belgium.	GC and LC standards
Alpha Analytical Laboratories Division of Alpha Metals, Inc. 56 Water Street Jersey City, NJ 07304, USA.	CRM, RM
Alltech Associates 2051 Waukegan Road Deerfield, IL 60015, USA.	GC standards, GC and LC packings
American Petroleum Institute Standard Reference Materials Carnegie-Mellon University Schenley Park, Pittsburgh, PA 15213, USA.	CRM, RM
Analabs, Inc. (Foxboro Analytical) 80 Republic Drive North Haven, CT 06473, USA.	GC standards, GC and LC packings
Applied Science Laboratories, Inc. P.O. Box 440 State College, PA 16801, USA.	GC standards, GC and LC packings, RM

Applied Science Europe B.V. P.O. Box 1149, 3260 AC Oud- Beijerland, The Netherlands.	GC standards, GC and LC packings, RM
Arro Laboratories, Inc. P.O. Box 686, Caton Farm Road, Joliet, IL 60434, USA.	CRM, RM
Association of Official Analytical Chemists, Box 540, Benjamin Franklin Station Washington, DC 20040, USA.	GC and LC standards
Baird Corporation 125 Middlesex Turnpike Bedford, MA 01730, USA.	CRM
Baird-Atomic Ltd. Warner Drive, Springwood Ind. Estate, Rayne Road, Braintree, Essex, Great Britain	CRM
J. T. Baker Chemical Company 222 Red School Lane Phillipsburg, NJ 08865, USA	CRM, RM, GC standards, GC and LC packings
Baker-Chemikalien, Postfach 1661, Gross-Geran German Federal Republic	CRM, RM, GC standards, GC and LC packings
BDH Chemicals, Ltd. Poole, Dorset BH12 4NN, Great Britain	GC standards
Brammer Standard Co. 213-Essex Knoll Drive Coraopolis, PA 19108, USA	CRM, RM
Brinkman Instruments, Inc. Cantiague Road Westbury, NY 11590, USA	RM
Burdick and Jackson Laboratories 1953 South Harvey Street Muskegon, MI 49442, USA	RM, GC and LC standards
Bureau of Analytical Samples, Ltd. Newham Hall, Newly, Middleborough TS8 9EA, Great Britain	GC and LC standards
California Bionuclear Corporation 7654 San Fernando Road Sun Valley, CA 91352, USA	CRM
Carlo Erba Strumentazione SpA P.O. Box 4342, 1-20100 Milan, Italy	GC standards and packings

Chem. Service, Inc. P. O. Box 194 West Chester, PA 19380, USA.	GC and LC standards
Chrompack Nederland BV P. O. Box 3, 4330 AA Middleburg The Netherlands,4330	GC standards, GC and LC packings
Coast Engineering Lab 13508 S. Normandie Avenue Gardena, CA 90249, USA.	GC supports
P. J. Cobert Associates, Inc. 10767 Indian Head Industrial Drive St. Louis, MO 63132, USA.	GC and LC packings
Columbia Scientific Industries 11950 Jollyville Road P.O. Box 9908, Austin, TX 78766, USA.	CRM, RM
C. Desaga GmbH, Nachf. Erich Fecht Maasstrasse 26-28, P.O. Box 101969, D-6900 Heidelberg 1 German Federal Republic	GC standards
Duke Scientific Corporation 445 Sherman Avenue, Palo Alto, CA 94306, USA	CRM, RM
E. I. du Pont de Nemours and Co. Analyical Instruments Division Room 38830 Wilmington, DE 19898, USA	LC packings
Du Pont U.K, Ltd. Analytical Instruments Wedgewood Way, Steuenage Herts, SGI 4QN, Great Britain	LC packings
E. I. du Pont de Nemours (Deutschland) GmbH Instrument Products Division Dieselstrasse 18, P.O. Box 1509, D6350 Bad Neuheimel, German Federal Republic	LC packings
Eastman Kodak Company Eastman Organic Chemicals Rochester, NY 14650, USA	CRM, RM
EM Laboratories, Inc. 500 Executive Boulevard Elmsford, NY 10523, USA	RM
Environmental Protection Agency Quality Assurance Branch Environmental Monitoring and Support Laboratory Cincinnati, OH 45268, USA	QC standards

Environmental Resource Associates Department A 120 East Sauk Trail South Chicago Heights, IL 60411, USA.	Environmental reference samples, potable water reference samples, pesticide reference samples
ES Industries 8 South Maple Avenue Marlton, NJ 08053, USA.	LC packings
Foxboro Analytical, see Analabs, Inc.	
Foxboro Nederland N.V. Analytical Division S. Gravelandseweg 557, P.O. Box 113 Schiedam, The Netherlands.	GC standards, GC and LC packings
Hopkins and Williams P.O. Box 1, Romford, Essex RM1 1HA, Great Britain.	GC and LC packings
Hewlett-Packard Company 150 Page Mill Road Palo Alto, CA 94304, USA.	GC packings, fused silica OTC
ICN Life Sciences Group 26201 Miles Road Cleveland, OH 44128, USA.	CRM, RM
Jarrel-Ash(Div. of Fisher Scientific Co.) Spectrographic Supplies Section 590 Lincoln Street, Route 128 Waltham, MA 02154, USA.	CRM, RM, GC standards, GC and LC packings
JASCO, Inc. 218 Bay Street, Easton, MD 21601, USA.	LC packings
Johns Manville Ken-Caryl Ranch, Denver, CO 80217, USA.	GC and LC packings
L.C. Company, Inc. 619 Estes Avenue Schaumberg, IL 60139, USA.	GC standards, GC and LC packings
Leco Corporation 3000 Lakeview Avenue, St. Joseph, MI 49085, USA.	CRM, RM
Macherey-Nagel and Company, GmbH, Export, Werkstrasse 6-8, P.O. Box 307 D-5160 Duren, G.F.R.	GC and LC packings
Matheson Company P.O. Box E Lyndhurst, NJ 07071, USA.	GC standards, GC and LC packings

Matheson Company Nijverheidstraat 23B, B-2431 Olevel, Belgium	GC standards, GC and LC packings
Nanogens-Analytical Specialists P.O. Box 1025, Watsonville, CA 95076, USA.	RM
Packard-Becker BV P.O. Box 519, Vulcanusweg 259 26244 AV, Delft, The Netherlands.	GC packings
Packard Instrument Company 2200 Warrenville Road Downers Grove, IL 60515, USA.	GC packings
N.V. Packard Instrument S.A. Lakenstraat 168 B-1000, Brussels, Belgium.	GC packings
Perkin-Elmer Corporation Main Avenue, Norwalk, CT 06851, USA.	GC and LC packings
Phase Separations Ltd. Deeside Industrial Estate Queenferry, Flintshire, Great Britain.	GC and LC packings
Phase Separations Ltd., 255 Oser Avenue, Hauppauge, NY 11787, USA.	GC and LC packings
Phillips Petroleum Co. Special Products Division Bartlesville, OK 74004, USA.	CRM, RM
Pierce Chemical Company P. O. Box 117, Rockford, IL 61105, USA.	GC and LC packing, standards
Pierce Eurochemie, B.V. Box 1151, Rotterdam, The Netherlands	GC and LC packings, standards
Polyscience Corporation 6366 Gross Point Road Niles, IL 60648, USA.	CRM, RM, GC standards
Pye Unicam, Ltd. York Street, Cambridge CBI 2PX, Great Britain	GC and LC packings
Regis Chemical Company 8210 N. Austin Avenue Morton Grove, IL 60053, USA.	GC standards

Research Organic/Inorganic Chemical Corp. 11686 Sheldon Street Sun Valley, CA 91352, USA.	CRM, RM
Scientific Gas Products, Inc. 513 Raritan Center, Edison, NJ 08817, USA.	CRM
The Separations Group P.O. Box 867, 16640 Spruce Street, Hesperia, CA 92345, USA.	LC packings
Shimadzu Corporation, Ltd. International Manufacturing Div. 14-5 Uchikanda 1-chome, Chryoda-kui, Tokyo 101, Japan	GC packings
Scott Specialty Gases Scott Environmental Technology,Inc, Plumsteadville, PA 18949, USA.	CRM, RM, Cross reference service collaborative gas analysis
Supelco, Inc. Supelco Park, P. O. Box 581 Bellefonte, PA 16823, USA.	GC and LC standards, GC and LC packings, RM
Supelco S.A. European Rte De Cligny 3,1299 Crans, Switzerland	GC and LC standards, GC and LC packings, RM
Theta Corporation, Chemical Division, P.O. Box 167, Media, PA 19063, USA.	GC standards
United States Geological Survey Water Resources Division U.S. Department of Interior Washington, DC, 20234, USA.	Ampoule water concentrates, prepared natural water standards
Union Carbide Corporation, Linde Division, P.O. Box 372 51 Cragwood Road, South Plainfield, NJ 07080, USA.	RM
U.S. Department of Commerce Office of Standard Reference Materials, Washington, DC 20234 USA.	CRM, RM, Trace elements in National Bureau of Standards water, gases for heavy duty vehicle emissions analysis
U.S. Department of Interior Bureau of Mines, Branch of Engineering, Box H, 4372, Herring Plaza, Amarillo, TX 79101, USA.	CRM, RM

Universal Scientific, Inc. P. O. Box 80402, 2070 Peachtree Industrial Court, Atlanta, GA 30341, USA.	GC standards and packings, LC packings
Varian Associates 611 Hansen Way Palo Alto, CA 94303, USA.	GC standards and packings, LC packings
Varian MAT GmbH Barkhausenstrasse 2, Postfach 1440 62, 2800 Bremen 12, G.F.R.	GC standards and packings, LC packings
V-Tech Corporation 16229 W. Ryerson Road, P.O. Box 183, New Berlin, WI 53151, USA.	GC standards
Waters Associates, Inc. 34 Maple Street, Milford, MA 01757, USA.	LC packings
Whatman, Inc. 9 Birdewell Place, Clifton, NJ 07014, USA.	LC packings
Whatman Ltd. Springfield Mill, Maidstone, Kent ME14 2LE, Great Britain	LC packings

*CRM = Certified reference materials; RM = General reference materials.

REFERENCES

1 G. R. Umbreit, Trace Analysis by Gas Chromatography, in R.L. Grob (Ed.), Modern Practice of Gas Chromatography, Wiley-Interscience, New York 1977, p. 368.
2 E. O. Oswald, P. W. Albro and J.D. McKinney, J. Chromatog., 98, (1974), 363.
3 H. P. Burchfield and E. E. Storrs, J. Chromatog. Sci., 13 (1975), 202.
4 W. Thornburg, Anal. Chem., 47 (1975) 157R.
5 B. F. Saltzman and J. E. Cuddeback, Anal. Chem., 47 (1975), 1R.
6 M. J. Fishman and D. E. Erdemann, Anal. Chem., 47 (1975) 334R.
7 L. Weil, Muenchner Beitr. Abwasser-Fisch-Flussbiol., 19, (1971) 191-200, C.A., 76 (1972) 37137v.
8 A. A. Rosen, Symp. Org. Matter Natur. Waters, 1968, pp. 359-367; C.A. 74, (1971), 6743r.
9 P. M. Giever, Particulate Matter Sampling and Sizing, in A.C. Stern (Ed.), Air Pollution, Academic Press, New York, 1976, p.6.
10 W. H. Durum, Characteristics of Water Resources, in L. L. Ciaccio (Ed.), Water and Water Pollution Handbook, Vol. 1, Marcel Dekker, New York, 1971, Ch. 1.
11 R. I. Larsen, C. W. Zimmer, D. A. Lynn and K. G. Blemel, J. Air Pollut, Contr. Ass., 17, (1967) 85.
12 W. J. Youden, Statistical Methods for Chemists, Wiley, New York, 1951.
13 American Society for Testing Materials, ASTM Manual on Quality Control of Materials, Special Technical Publication, 15-C, ASTM, Philadelphia, PA, 1951.
14 E. L. Bauer, A Statistical Manual for Chemists, 2nd ed., Academic Press, New York, 1971.
15 American Society for Testing and Materials, Proposed procedure for determination of precision of committee D-19 methods, in Manual on Industrial Water and Industrial Waste Water, 2nd ed., ASTM, Philadelphia, PA, 1966.

16 J. D. Hinchen, Practical Statistics for Chemical Research, Methuen and Co., London, 1969.
17 A. J. Duncan, Quality Control and Industrial Statistics, 3rd ed., R. D. Irwin, Homewood, IL, 1965.
18 D. J. Cowden, Statistical Methods in Quality Control, Prentice-Hall, Englewood Cliffs, NJ, 1957.
19 A. L. Linch, Evaluation of Ambient Air Quality by Personnel Monitoring, CRC Press, Cleveland, OH, 1974, p.213.
20 J. D. Rhoades and L. Bernstein, Chemical, Physical and Biological Characteristics of Irrigation and Soil Water, in L. L. Ciaccio (Ed.), Water and Water Pollution Handbook, Vol. 1, Marcel Dekker, New York, 1971, Ch. 3.
21 D. Carroll, U. S. Geol. Surv. Water-Supply Paper. 1535G, U.S. Dept. Interior, Washington, DC 1962.
22 F. A. Herman and E. Gorham, Tellus, 9 (1957) 180.
23 G. H. Neumann, S. Forselius and L. Wahlman, Intern. J. Air Pollution, 2 (1959) 132.
24 G. A. Wheatley and J. A. Hardman, Nature, 207 (1965) 486.
25 S. R. Weibel, R. B. Weidner, A. G. Christianson and R. J. Anderson, Advances in Water Pollution Research, Water Pollution Control Federation, Washington, DC, 1967, pp 329-342.
26 Anonymous, Chem. Eng. News, Feb. 5 (1979) 17.
27 American Industrial Hygiene Association, Air Pollution Manual, Part I, Evaluation, 2nd ed., Detroit, MI, 1972.
28 M. W. First, Environ. Res., 2(2) (1969) 88-92.
29 H. J. Paulus and R. W. Thron, Stack Sampling, in A. C. Stern (Ed.), Air Pollution, Vol. III, 1976, pp 525-587.
30 J. S. Nader, Source Monitoring in A. C. Stern (Ed.), pp 589-601.
31 H.B.N. Hynes, The Biology of Polluted Waters, Liverpool University Press, Liverpool, 1960.
32 L. Klein (Ed.) River Pollution, II, Causes and Effects, Butterworths, London, 1962.
33 K. M. Mackenthum and W. M. Ingram, Biological Associated Problems in Freshwater Environments - Their Identification, Investigation and Control, Federal Water Pollution Control Admin., Washington, DC,1967.
34 C. N. Sawyer and P. L. McCarty, Chemistry for Sanitary Engineers, 2nd ed., McGraw-Hill, New York, 1967.
35 L. E. Keup, W. M. Ingram and K. M. Mackenthum, Biology of Water Pollution: A Collection of Selected Papers on Stream Pollution, Waste Water and Water Treatment, Publ. No. CWA-3, Federal Water Pollution Control Admin., Washington, DC,1967.
36 P. S. Welch, Limnology, 2nd ed., McGraw-Hill, New York, 1952.
37 G. E. Hutchinson, A Treatise on Limnology, Vol. 1, Geography, Physics, and Chemistry, Wiley, New York,1957.
38 G. E. Hutchinson, A Treatise on Limnology, Vol. 2, Introduction to Lake Biology and the Limnoplankton, Wiley, New York, 1967.
39 F. Ruttner, Fundamentals of Limnology (transl. by D. G. Frey and F. E. J. Fry), University of Toronto Press, Toronto, 1953.
40 G. K. Reid, Ecology of Inland Waters and Estuaries, Reinhold, New York, 1961.
41 Standard Methods for the Examination of Water and Wastewater, 12th ed., American Public Health Association, New York, 1965.
42 P. S. Welch, Limnological Methods, Blakiston, Philadelphia, PA 1948.
43 H. A. Painter, Chemical, Physical and Biological Characteristics of Wastes and Waste Effluents, in L. L. Ciccio (Ed.), Water and Water Pollution Handbook, Vol. 1, Marcel Dekker, New York, 1971, Ch. 7.
44 A. P. Altshuller, Analytical Problems in Air Pollution Control, Proc. Environ. Qual. Sensor Workshop, Las Vegas, NV, EPA, Washington, DC 1971, pp 11-44.
45 L. A. Rittmiller, B. M. Zonkowski, and I. L. Wadehra, Pollut. Eng., 3 (1971) 26-28.
46 Y. Saijo, Bunseki Kagaku, 21 (1972) 395-402; C.A., 76 (1972) 15808f.
47 D. L. King, Water Pollut. Handbook, 2 (1971) 451-481.

48 K. Grasshoff, Chem. Oceanogr., (1971) 68-69; C.A., 76 (1972) 37207t.
49 R. C. Kroner and D. G. Ballinger, Treatise Anal. Chem., Vol. 2 (1971) 343-412.
50 North Atlantic Treaty Organization/Committee on Challenges to Modern Society, Brussels, Belgium. Air Quality Criteria, Sulfur Oxides (N-7), Particulate Matter (N-8), Carbon Monoxide(N-10), Nitrogen Oxides (N-15), Photochemical Oxidants and Related Hydrocarbons (N-29), U. S. Government Environmental Protection Agency, Research Triangle Park, NC, 1974.
51 Division of Medical Sciences, Biological Effects of Atmospheric Pollutants, Fluorides ISBN-0-309-01922-2(1971); Asbestos ISBN-0-309-01927-3, Lead ISBN-0-309-01941-9, and Particulate Organic Matter ISBN-0-309-02027-1 (1972); Medical and Biological Effects of Environmental Pollutants, Manganese ISBN-0-309-02143-X(1973), Chromium ISBN-0-309-02217-7, Vanadium ISBN-0-309-02218-5, and Nickel ISBN-0-309-02314-9(1974). National Research Council-National Academy of Sciences, Washington, DC, Vapor Phase Organic Pollutants-Volatile Hydrocarbons and Oxidation Products (NTIS-PB 249 357) (1975), Arsenic (NTIS-PB 262 167) (1976), Chlorine and Hydrogen (NTIS-PB 253 96), Ozone and other Photochemical Oxidants (PB 260 570-571, Selenium (NTIS-PB 251 318) (1976).
52 Advisory Center on Toxicology, Short-Term Limit Reports, Oxides of Nitrogen PB-199 903, Basis for Guides PB-199 904, Hydrogen Chloride PB-203 464, and Hydrogen Fluoride PB-203 465 (1971); Ammonia PB-244 336 (1972); Carbon Monoxide PB-244 338 and Chlorine PB-244 339 (1973); and Hydrazone, Monomethyl Hydrazine, and 1,1-dimethylhydrazine PB-244 337 (1974). National Research Council-National Academy of Sciences, Washington, DC.
53 World Health Organization, Air Quality Criteria and Guide for Urban Air Pollutants, Tech. Rep. Ser. No. 506, World Health Organization, Geneva 1972.
54 World Health Organization Environmental Health Criteria Programme, Report of a Meeting on Environmental Health Criteria and Standards, Geneva 20-24 November 1972, EP 73/1, World Health Organization Geneva, 1973.
55 Air Quality Abstracts, Pollution Abstracts, LaJolla, CA, 1974.
56 Air Pollution Technical Information Center (APTIC) Air Pollution Abstracts, U.S. Gov't. Printing Office, Washington, DC.
57 D. L. Brenchley, C. D Turley and R. F. Yarmac, Industrial Source Sampling, Ann Arbor Sci. Publ., Ann Arbor, MI, 1973.
58 M. Sittig, Pollution Detection and Monitoring Handbook, Noyes Data Corp., Park Ridge, NJ, 1974.
59 T. Cercer, Aerosol Technology in Hazard Evaluation, Academic Press, New York, 1976.
60 Environmental Instrumentation Group, Instrumentation for Environmental Monitoring Air, 1st ed., Lawrence Berkeley Laboratory, University of California, Berkeley, CA, May 1972; updates: Feb. 1973, Dec. 1973.
61 Air Sampling Instruments Committee, ACGIH, Air Sampling Instruments, 4th ed., American Conference of Governmental Industrial Hygienists, Cincinnati, OH, 1972.

Chapter 3

SAMPLING TECHNIQUES

3.1 INTRODUCTION

Sampling is the operation of removing a part (of convenient size) from the whole in such a manner that the sample represents, within measurable limits of error, the proportion of the quantity in the whole.

The principles of sampling are of importance to the analytical chemist for several reasons:

(1) The greatest care taken in performance of the analysis may be in vain if the sample has been obtained carelessly; i.e., it is not representative of the system being sampled.

(2) The proper method of treating the sample(s) is dependent upon the purpose of the sampling.

Methods of analysis are more precise than most sampling procedures. Through experimentation it is possible to determine the contribution of errors of analysis and sampling to the total variability of results and to control the combined procedure accordingly. An economic balance should be maintained between the number of subdivisions of a lot (the bulk sample) of material that are sampled and the number of subsamples from each subdivision.

To perform the job properly, one should know as much as possible about the nature and condition of materials to be sampled, the tools to be used in the sampling and have adequate knowledge about analytical methodology. One of the purposes of sampling is to judge acceptability of the sample. Most often, a sample is taken because one wishes to know if the material it represents meets certain specifications. For this purpose, the sample must accurately represent the whole quantity (the population available) under consideration.

Another facet of the acceptability is to assure the absence of contamination or dirt from material being sampled. For this important purpose, it is usually preferable to sample in a way that gives maximum assurance of finding contamination, if the contamination present does not accurately represent the bulk quantity.

A third purpose of sampling is to identify unknown materials. The expense of obtaining an accurate sample may not be warranted because a particular contamination does not persist long enough to establish its identity.

Sampling techniques for environmental samples include the three states of matter: gas, liquid and solid. We will survey each of these in reference to water and air pollution.

A number of restrictions are placed on all sampling methods. The samples collected must represent the conditions at the time of sampling. To permit accuracy of testing the samples must be of sufficient volume and number to fulfill the reasons for the sampling. The samples must be collected, labeled, packed, shipped and manipulated prior to laboratory treatment in such a way that these processes minimize any changes in a particular component of interest or sample properties to be examined.

The time interval between sampling and analysis should be limited for a number of reasons:

(A) decrease adsorption of gases and solids from sample to walls of the container;

(B) minimize sample loss because of leaks;

(C) decrease possibility of chemical changes because of interactions of sample components or photochemical decomposition.

The philosophy behind this time restriction is that a minimum time interval between sampling and analysis decreases the chances for any change in the sample (physical or chemical).

3.2 BACKGROUND AND GENERAL DISCUSSION

A number of general references are available to the person desiring additional information on this very important aspect of environmental analysis (refs. 1-13).

The volume of sample collected should be held to a minimum, its size being controlled by the volume needed for an accurate analysis. It is not wise to choose the sample volume by the lower limits of precision and accuracy, but rather to choose a volume at least 2.5 times this optimum value. The number of samples taken should be sufficient so that one may assume a representative sampling has been completed. This is not always possible as the time (to sample and analyze) and cost are usually important parameters. A compromise must be made especially when sampling from a lake, reservoir, etc., where a single sample will permit minimum information. If the sample site (e.g., lake or other large body of water) is fed by smaller streams, then it becomes important to sample above and below the point of entrance to the large site. The non-movement of the water in a lake makes it more difficult to point out waste contamination by sampling at one site than if the sample were taken from a flowing stream. Measurement of the concentration and volume of waste material in waste streams is very important. Concentrations of the pollutants in waste effluents are not useful alone; they must be interpreted in terms of interactions with the receiving waters. Thus, flow rates of both the waste streams and the receiving waters must be known.

Contamination of water results from three broad groupings: dissolved gases, dissolved liquids and solids, and particulate solids. Oxygen is probably the most commonly dissolved gas measured in aquatic environments. If one is interested in BOD (biological oxygen demand) or COD (chemical oxygen demand) values, then the concentration of dissolved oxygen should be determined at least twice each sampling day - at midafternoon and very early in the morning before the sun rises. Carbon dioxide is another gas of considerable interest for aquatic studies. Its importance arises because of the carbonate-bicarbonate buffer system in all water systems. However, the analysis of free carbon dioxide is not a useful endeavor because of this buffering system which is present.

Dissolved solid substances are defined as those which pass through a 0.45 µm filter. Nitrogen and phosphorus nutrients, chloride, sulfate and carbonate salts, and heavy metals and cyanide (toxic materials) are the predominately dissolved ionic inorganic solids. With the exception of the carbonate-bicarbonate buffer system, which serves as a reserve carbon source for photosynthesis (ref. 14), these inorganics do not show much diurnal variation (ref. 15).

Dissolved organic materials generally may be classified as:

(1) those dissolved nontoxic organics present in waste waters. These are an energy source for the heterotrophic microbiota both in the receiving waters and the waste waters. They are referred to as biodegradable organic substances.

(2) those dissolved organics which are formed during anaerobic fermentation of organic wastes (ref. 13). These are not toxic to nor used by aquatic biota. This type of dissolved organic material is responsibile for taste and odor and usually may be collected on carbon filters.

(3) those dissolved organics which are toxic to aquatic biota and which may be found in natural as well as waste waters, e.g., organic pesticides.

The dissolved organic materials are usually present in much lower concentrations than inorganics, mainly because of their low solubility in water. Due to their low concentration, more precise and sophisticated analytical techniques are needed for identification and measurement, e.g., extractions, chromatographic analysis and spectroscopy. The inorganic fraction of particulate solids is mainly silt, sand and clay soil particles, but the organic fraction may be either living or dead organic materials.

Sampling volatile constituents of contaminated atmospheres requires efficient collection, concentration of components, and removal of sample for analysis. Most analyses of this type deal with trace quantities of impurities, and for this reason, large samples are usually necessary (ref. 16). Density of gaseous pollutants (e.g., phosgene, mustard gas, lewisite or chloride) usually counteract diffusion processes to the extent that it may cause stratification (ref. 17). It is for this reason that one should not depend upon diffusion processes to produce homogeneous mixtures in most cases.

There are a number of effects which can result in concentration errors in many sampling techniques (ref. 18). (1) adsorption on walls and connecting tubes, (2) solid adsorbent collection system effects, (3) diffusion through plastics, (4) mechanical defects in equipment, (5) partial vapor pressure effects, (6) solubility effects, (7) temperature effects, (8) volumetric errors, (9) errors of observation, and (10) errors in sampling rate.

Most sampling systems for airborne pollutants usually consist of four components: (1) intake and transfer section; (2) collection device; (3) flow measuring component, and (4) component to keep air moving through the system.

Each of these components must perform properly for a successful sampling scheme. One may consider the sampling system as the receptor for the component or effect to be measured. Any subsequent changes considered must be done with referral to the receptor and what consequences the receptor may have upon the success of the sampling.

Ambient air quality is influenced by the various meterological parameters existing at the time of the sampling (e.g., stability of the atmosphere, precipitation, wind velocity and direction, humidity, temperature and solar radiation). Temperature decreases as one goes higher in altitude. This is referred to as lapse rate; and, for dry adiabatic air, it is 9.8° C/1000 m (ref. 19). The levels of air pollution at ambient conditions have been found to be inversely related to the wind velocity (ref. 20).

When choosing a sampling site for continuous monitoring, it is necessary to consider the proper siting of the sampling system to provide proper representation and to consider the sites in reference to each other (refs. 21-25). Items which should be considered when selecting a sampling site are: (a) the sampling inlets should not be lower than 2 m above ground level to minimize collection of contaminants at Earth's surface, (b) sampling sites should not be located on the side protected from the wind by tall buildings, (c) the sampling site should be easily accessible and secure from tampering, (d) the intake of the sampler should not be exposed directly to localized pollution sources, (e.g., chimney), and (e) the site should have adequate electrical power for maintaining the equipment.

Two terms commonly found in the environmental sampling literature are source sampling and emission sampling. These two terms are very often used interchangeably, however, they mean different things. Source sampling should be considered as the sampling of the environmental pollutants at their source, i.e., before they have a chance to enter the atmosphere and become diluted. On the other hand, emission source sampling refers to the sampling of the pollutants once they have entered the atmosphere and are diluted. Emission sources are usually monitored to determine mass emission rate of certain pollutants from a source, to obtain emission data from a number of sources for input data to an air quality model and to evaluate control devices which have been installed at certain sources of pollution.

The sampling and analysis of air and water is of interest because it concerns a topic which is contemporary and it uses techniques and principles which may or may not be familiar to most scientists. New environmental restrictions are continually being enacted and old ones being relaxed. Environmental air problems have to include all volumes of air starting at the surface of the Earth to a distance at least 7 miles above the Earth's surface. This gaseous system is commonly known as air and should have the composition (whether it be indoors or out-of-doors): N_2 (78.08%), O_2 (20.95%), CO_2 (0.034%), Ar (0.93%), and smaller percentages of H_2, Ne, Kr, Xe, and He. Other components can be classified as contaminants. In the category of contaminants we find such things as hydrocarbons, carbon monoxide, sulfur compounds, nitrogen compounds, metals, metal ions, various oxidants, pesticides, etc. As has been pointed out by Axelrod and Lodge (ref. 26), "pure air" presents a problem of definition because even the purest air that we can sample on earth, in the natural state, will contain hundreds of contaminants. Thus, the best we can do is to use an operational definition: "pure air" is then the air which is free of anything known to interfere with the analysis. Environmental water problems have to include any material(s) which may be injurious to the well-being of mankind. Here, water problems have various orders of magnitude, depending upon which water we are discussing and for what use it is intended. The problems of odors emanating from water supplies enters the area of both water and air pollution. If the purpose of the sampling and analysis is to determine whether the source of the odor is in the water, then it is a water pollution problem. If the purpose is just to identify the odor, then it could rightly be classified as an air pollution problem.

In the U.S.A., air and water quality standards have been issued by the Enivironmental Protection Agency (EPA) (refs. 27 and 28). These are all in need of updating and other contaminants need to be added to the list. Enclosed and/or working areas have had restrictions placed on them by OSHA (Occupational Safety and Health Administration) and NIOSH (National Institute of Occupational Safety and Health). In Chapter 8, the authors have tabulated all standards throughout the world which are known or available at the time of this printing.

3.2.1 Theory and statistical foundation of sampling

A fitting way to start this section would be to give a direct quote from a chapter on trace analysis by gas chromatography written by Umbreit (ref. 29). "In this society we have reached the point of enacting laws providing rules limiting at trace levels the discharge of a wide variety of specific compounds into our environment and have provided penalties for exceeding such limits. As a consequence, even though there is reasonable uncertainty in the identification of specific compounds at trace levels, regulatory action is taken on the basis of analytical measurements, in many cases without regard to an evaluation of the accuracy of these measurements." The authors feel the passage speaks for itself and no further explanation is necessary.

The justification for sampling is quite an easy task if one considers analyzing all possible air/and water environments on a continuing basis. The time and cost, however, are limiting factors to such a project. So, before we begin a sampling program we should answer a number of questions. Is our objective to secure data about every air and water population (population meaning every unit which goes to make up a body of water, the outside air of the environment, confined air environments, e.g., work areas or a hospital) or is it only necessary to obtain exact information of representative parts of these populations? Also, do we need this information on a continuing basis or for a specified period of time? In other words, do we want to know if a level of toxicity has been exceeded, or do we wish to know what the level of contamination is at a particular time and location. What these questions come down to are: for what is the data to be used and how accurate does it have to be? One must realize that a certain level of error will be present in the sampling processes as well as in the measuring processes. The degree of accuracy of the sampling and measuring steps should be compatible. By that we mean, do not use a very elaborate sampling scheme when your measuring step is accurate enough only to give a yes or no answer as to whether a certain level has been exceeded. Conversely, one would not use an unplanned sampling procedure when the measuring step must have a very high degree of accuracy and precision. Conclusions drawn about the population from a sample may be completely different from the actual population values. This difference is mainly dependent upon the sampling survey rather than on the analyses of the samples.

Sampling error usually decreases with an increase in sample size (i.e., the number of units n selected in the sampling). In fact, in many situations the decrease is inversely proportional to the square root of n (sample size). Sampling error is the error arising from drawing inferences about the population (bulk sample) on the basis of observations on a part of the population (the samples). Fig. 3.1 shows that sampling error decreases substantially as n increases, however, increasing sample rate becomes a marginal consideration as n continues to increase. A similar comment may be said for the reliability of the arithmetic means, $\bar{X}$, as compared to the number of measurements. In other words, the reliability of the mean increases as the square root of n (number of measurements).

The mean is 3 times as reliable as that of a single measurement when n=9;

The mean is 4 times as reliable as that of a single measurement when n=16, etc.

Let us look at the sampling process on a more quantitative basis. Suppose we wish to analyze a body of water (e.g., a lake) for the concentration of the insecticide 2,4-D. If we sampled the entire lake, our population sampling units

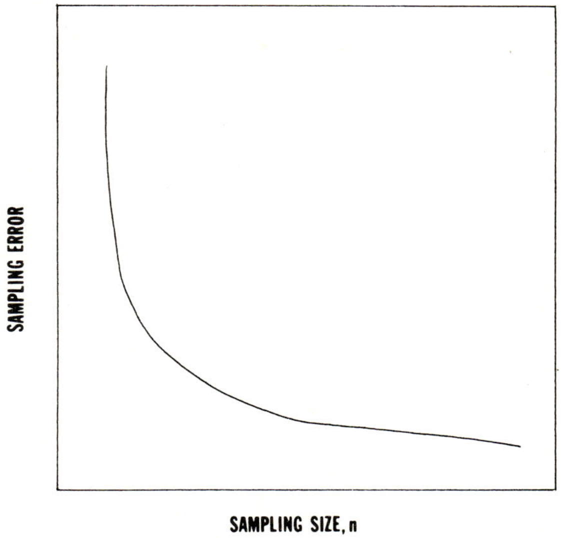

Fig. 3.1 Sampling error change with an increase in number of samples.

would be N (a population). Analysis of all these samples would provide an error, of $\%E_N$. This error would be due to the presence of non-sampling errors because we sampled and analyzed the entire population. If we selected a random set of samples, n, from the population, N, the population error resulting from this finite set of samples (i.e., the divergence from the true value, μ) would be $\%E_n$. $\%E_n$ would be composed of two types of error: sample error, $\%E_{n1}$, and non-sampling error, $\%E_{n2}$. We may now define some other terms:

C_N = cost of collecting and analyzing data for each sample from the population (N).

C_n = cost of collecting and analyzing data for each sample in the finite set of samples (n).

Now C_N should be greater than C_n; however, nC_n should be less than NC_N.

In many cases, the sampling and non-sampling errors, $\%E_n$, in the sample survey for a correct size, n, will probably be less than the non-sampling errors,

$\%E_N$. The total cost for any given sample size can be written:

$$nC_n + LE_{n1} + LE_{n2} \qquad (3.1)$$

where L = loss involved in making decisions on the survey results, i.e., loss per 1% error.

So as n increases, the components nC_n and LE_{n2} would increase, whereas, LE_{n1}, would decrease. The value of the sample size n where the total cost reaches a minimum, is considered the optimum sample size n. This cost will go through a minimum (see Fig. 3.2).

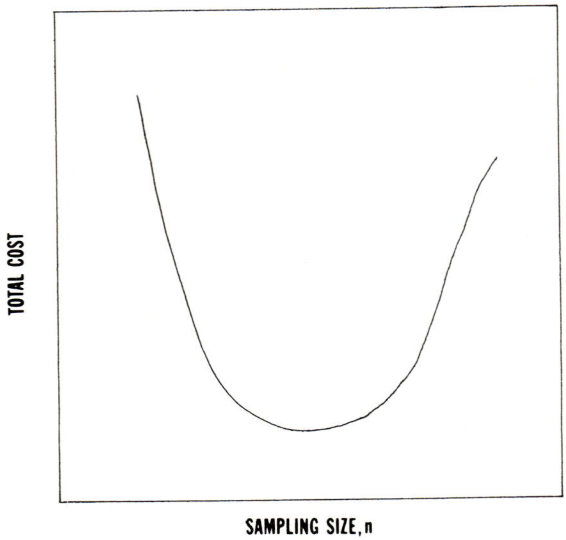

Fig. 3.2 Relationship of sampling cost compared to number of samples taken.

The relative standard error of the sample mean $\bar{X}$ from the measurement of n samples collected by simple random sampling without replacement is given by

$$C(\bar{X}) = [(N-n)/(N-1)]^{1/2} C/(n)^{1/2} \qquad (3.2)$$

where C = population coefficient of variation.

To insure a relative standard error of %e, we need a sample size, n, of

$$n = NC^2/[(N-1)e^2 + C^2] \qquad (3.3)$$

A random sample is one whose selection is based on the theory of randomness, i.e., it is a sample drawn in such a manner that each unit in the population had an equal predetermined probability of being selected. Thus, a probability sampling scheme would be one in which: (1) each part (individual sample from the population) has an equal probabiity of being taken, (2) these discrete samples are chosen by a random process that is consistent with the probabilities, and (3) the analysis of these discrete samples is based upon weights consistent with the probabilities set in step one (ref. 30).

If we sample n parts, with equal probability, the number of possible samples is $\binom{N}{n}$ and the selection probability of each sample is $1/\binom{N}{n}$. The cost function of a sampling program, C, can be estimated by (ref. 31)

$$C = C_o + nC_1 \qquad (3.4)$$

where C_o = the overhead costs for the program

C_1 = the average cost of analyzing one sample unit

n = the number of sample units in the program.

Laboratory samples for quantitative analysis are assumed to be representative of the population lot from which they were drawn. For uniform samples such as samples from a clear solution, any one portion would be a representative sample. Environmental samples, on the whole, are collected from a non-uniform population; so the usual practice is to select a number of portions to constitute the sample on the basis that the average value of the component in the sample will be essentially the same as the average that would be found if the entire lot had been analyzed. Rules of thumb from past experience and intuition are often relied upon for these purposes. A probability sample would furnish criteria for estimating the sample size requirement and precision to be achieved with more reliability.

A probability sampling plan makes use of the theory of probability to select sample portions or items with appropriate procedures for summarizing the test results. In this way a valid objective evaluation may be made of the probable limits of sample error.

A population material may be conceived as composed of N distinct portions of units, each possessing the given property but not necessarily to the same extent or magnitude. The units may be individual items or small quantitites taken from a population by means of a sampling device. Any group of n units could be considered a sample of this universe if the n units have been independently selected at random. Such a sample would permit certain inferences to be drawn; and, from these inferences, one could answer a variety of questions about the population.

If the sampling is performed correctly, a second probability sample of n units drawn from the same population should have an average value or range of values essentially of the same order of magnitude.

An important part of a sampling plan is the determination of the number of samples needed so sample averages will have the required precision. The calculation of minimum sample size requires an advance estimate of the standard deviation of the units in the sample population. A fairly good estimate may be derived from the range (lowest to highest value in the sample units) of the property to be determined and the shape of the distribution curve.

(a) If data values run rather uniformly throughout the range, w, the standard deviation, s, will be about 0.3w.

(b) If the distribution of data values is skewed (lying at one end or the other), then the standard deviation, s will be about 0.25w.

(c) If the data values are predominantly in the middle, the standard deviation, s, is about 0.2w.

A level of precision must be specified, i.e., we must decide: (1) what difference can be tolerated between the estimate to be made from the sample and the result that would be obtained by testing every unit in the population, and (2) what risk is acceptable if the tolerance is exceeded ? Keep in mind that the smaller the tolerance or the risk, the larger will be the sample size required.

Most evironmental samples may be considered single-stage sampling. If the universe consists of N units, n of which may be drawn as samples, then (ref. 29)

$$n = (t\sigma/E)^2 \qquad (3.5)$$

where $n << N$

t = factor corresponding to the acceptable risk of exceeding E;

σ = advance estimate of the standard deviation;

E = difference that can be tolerated between the sample estimate and the universe.

There are a number of values for t, depending on the risk we wish to take. These are shown in Table 3.1.

TABLE 3.1

Factor values corresponding to the risk of exceeding difference chosen for sample estimate and the value of the population (ref. 32)

t	Probability that E will be exceeded
3	3 in 1000
2.58	1 in 100
2	45 in 1000
1.64	1 in 10

When N is not large compared to n, then a limiting value, n_L must be calculated.

$$n_L = Nm/(N + n) \qquad (3.6)$$

If the sampling is based on an estimate of σ, once sampling and analyses have been completed a value of E, based on experimental data may be calculated by combining eqns. 3.5 and 3.6.

$$E = t\sigma[(N-n)/nN]^{1/2} \qquad (3.7)$$

The new value of E is independent of any error in the advance estimate (refs. 33-35).

3.2.1.1 Significance and rejection of data. Assuming the sampling phase of the environmental task has been completed successfully, one proceeds to the analysis of the samples. Many people undertaking analysis of environmental samples (or any type of sample for that matter) are blindly led by the belief that, if the analytical technique employed for the analysis possesses a relatively high degree of accuracy and precision, any stray outlying pieces of data may be discarded. Nothing could be further from the truth. Unless the person performing the analysis knows of some bona fide reason that the sample was not representative or that some step in the procedure was not followed, an outlying piece of data cannot be rejected. If one of the replicate runs for the separation and analysis of the environmental sample by either gas or liquid chromatography appears different from the others - mark the chromatogram accordingly. If this problem occurs, repeat the separation and analysis of the sample and mark this repeat chromatogram accordingly.

After the calculations are complete and one still has an outlying result, he may check his set of data for rejection of that piece of data. There are a number of these methods available to the analyst. He may check the outlying data by means of the Q-test (refs. 36 and 37) or the validity test using Student's t-tables (ref. 38). The Q-test is an easy method for checking from 3 up to 25 pieces of data.

To use the Q-test arrange the data in numerical sequence and determine the range w of the set of data.

$$\text{Range} = w = x_n - x_1 \qquad (3.8)$$

The end pairs of the set of data may then be checked:

(a) Smallest value (X_1)

$$Q = (X_2-X_1)/(X_n-X_1) = (X_2-X_1)/w \quad (3.9)$$

(b) Largest value (X_n)

$$Q = (X_n-X_{n-1})/(X_n-X_1) = (X_n-X_{n-1})/w \quad (3.10)$$

If Q calculated $\geq$ Q table, the piece of data may be rejected at the confidence level chosen.

To use the validity test based on Student's t-tables, one needs to have calculated the standard deviation, s, of the set of data.

$$s = \sum_{i=1}^{n} \left[(X_i-\bar{X})^2/(n-1) \right]^{1/2} \quad (3.11)$$

From a table of t-values one chooses the risk (α) that the true value (μ) will fall outside of the confidence limits if an infinite number of measurements had been made. The confidence limits (C.L.) are calculated from

$$C.L. = \bar{X} \pm (ts/(n)^{1/2}) \quad (3.12)$$

If the risk (α) is 0.10, then 0.10 is the area under the normal distribution curve that one would expect the true value (μ) to fall outside the confidence limits. In other words, 0.05 (5%) on either tail of the curve. Another way of interpreting this is that one feels there is a 90% probability that the true value, μ, will fall within the calculated confidence limits. Many analytical chemists choose a risk (α) of 0.05, which means a confidence limit of 95% or approximately two standard deviations.

3.2.1.2 Reliability and/or confidence in analytical data. Once the outlying results of a set of analytical data have been checked by: (a) legitimate physical reasons for elimination (e.g., column and/or detector overload because of too large a sample). A recent publication discusses this very important topic at a level anyone using chromatography can understand (ref. 39), (b) the Q-test, and (c) validity test using student's t-values, those remaining may be used to calculate the average value of the sample, $\bar{X}$, and the standard deviation, s. If a standard sample had been analyzed to check operator technique, variation in instrumental conditions, repeatability of sample delivery, etc. (a routine procedure that should always be followed), the data can be used to calculate the relative error.

Relative error (0/00)=[(X-μ)/μ]x1000 (3.13)

3.2.2 Aerosols

The collection of an aerosol sample is often difficult and requires attention to details. By an aerosol we mean the general classification of lyophobic (solvent-non-affinity) colloids. This class of colloids includes all systems in which the dispersed material is not in true solution. Although lyophobic systems are unstable, they may persist unchanged for long periods of time. The aerosol is a more specific classification of the lyophobic colloid which may be of two types: a smoke which is a solid dispersed in a gas, or a fog which is a liquid dispersed in a gas. Aerosols are generally formed by condensation (i.e., aggregation) of the dispersed phase in the matrix phase.

If aerosol formation is suspected, there are a number of reasons why one would be interested in their collection (sampling):

(1) to decide whether or not a hazardous concentration of a pollutant is present in the atmosphere and, more especially, if ambient air standards have been exceeded;

(2) to determine the effectiveness of monitoring programs and to reduce ambient concentrations of pollutants;

(3) to estimate emission levels at a source;

(4) to judge the effectiveness of control equipment;

(5) to pinpoint pollution sources.

Tobacco smoke is an example of a liquid aerosol, as is forest-fire smoke. "The science and technology...is...advanced to the point where aerosols of many compositions and particle sizes can be produced at will and their characteristics measured quite accurately (ref. 40)". Aerosols in the atmosphere vary in concentration and in time. The particles range from a few hundredths to a few tenths of a micrometer.

Stratospheric aerosols have been found to consist mainly of sulfate ion (SO_4^{2-}), small amounts of ammonium and hydrogen ions (NH_4^+ and H^+) and water. The stratosphere is the layer just above the troposphere which is the lowest layer of the atmosphere.

Tropospheric aerosols have sulfate ion contents of a few percent to nearly 100% (w/w). Tropospheric aerosols may also contain nitrate ion (NO_3^-), chloride ion (Cl^-), silicates, ammonium ions (NH_4^+), sodium ions (Na^+), calcium ions (Ca^{2+}) and traces of metals such as iron, lead, vanadium, manganese and magnesium. An appreciable quantity of organic material is also found in tropospheric aerosols. At one time or another, there are large aerosol particles (>1μm) in the troposphere. A negligible fraction of aerosol particles is in normal stratospheric air.

The variation in particle size may be due to the shape of the particle (which will affect its aerodynamic properties), the density and velocity (which affects inertia and electrical charges).

The momentum of aerosol particules (product of its mass and velocity) is a source of error in aerosol sampling. Aerosol particles are larger than gas molecules and each time the flow direction changes the larger particles continue on their original line and are displaced somewhat from the original part of the gas stream. The problem becomes more severe as particle size and velocity increase. Thus, it is necessary for all parts of the stream to be sampled and properly weighted so that a representative sample is obtained.

To insure that there is no change in momentum and that the sample will be representative of gases, as well as aerosols, one samples isokinetically. This is accomplished by using a thin-walled tube aligned with the stream flow and drawing sample into it at the same linear velocity as the stream flow at that point. Reasons for failure to obtain a truly representative sample may be related to a number of factors:

(1) The air flow is always disturbed by the presence of the probe.

(2) The sampling point chosen may not be representative of the whole stream.

(3) Turbulent flow could be set up at the point of sampling with the result that the sample will contain more of the smaller particles than the larger particles.

(4) Sample may be lost in the sample line by deposition, agglomeration or dispersion.

Studies have shown that (ref. 41):

(a) Isokinetic sampling is dependent upon sample alignment as well as velocity.

(b) Isokinetic conditions favor the collection of all particle sizes.

(c) The small mass of the smaller particles (under 3 μm diameter) minimizes inertial effects.

(d) Slow-moving particles have little or no momentum, thus ambient air does not require isokinetic conditions.

The processes taking place in Fig. 3.3 are the following:

(1) Fig. 3.3a shows isokinetic sampling, i.e., collection of an amount of particles proportional to what actually exists in the stream. Thus $V_s = V_n$ and $C_m = C_t$.

Fig. 3.3 illustrates the effect of non-isokinetic sampling as compared to isokinetic sampling of gases (with or without aerosols or particulates). Let us define some terms and symbols.

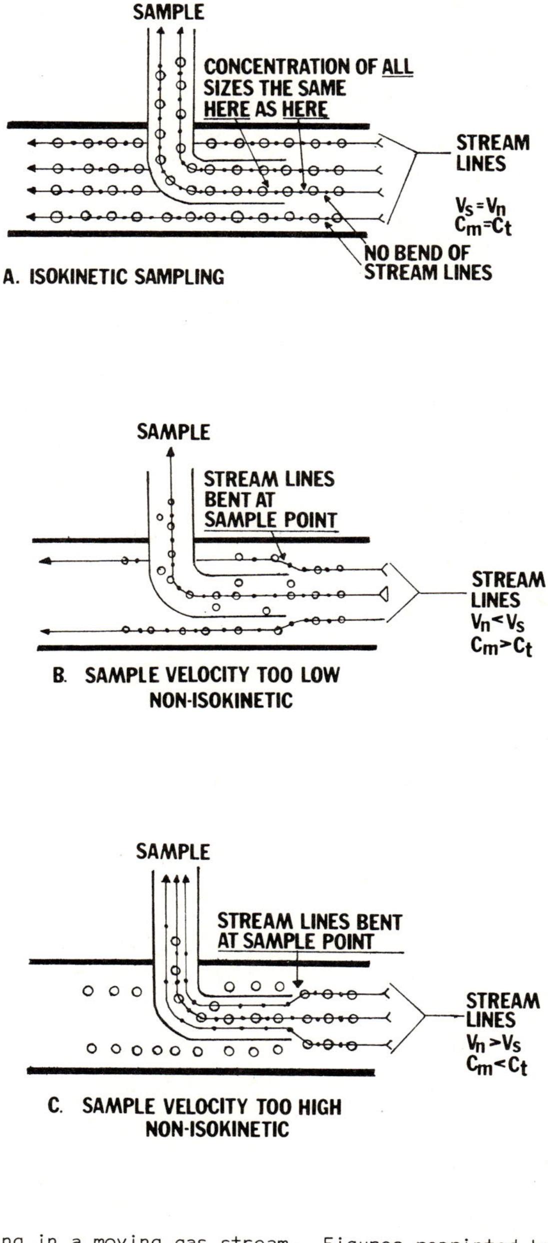

Fig. 3.3 Aerosol sampling in a moving gas stream. Figures reprinted by permission of the American Public Health Association from Methods of Air Sampling and Analysis, second edition, 1977.

V_n = velocity of the gas stream, i.e., sampling velocity;
V_s = velocity of the stack or duct gas;
C_m = measured concentration of the pollutant in the gas stream;
C_t = "true" concentration of the pollutant in the gas stream.

The processes taking place in Fig. 3.3 are the following:

(1) Fig. 3.3a shows isokinetic sampling, i.e., collection of an amount of particles proportional to what actually exists in the stream. Thus $V_s = V_n$ and $C_m = C_t$.

(2) In Fig. 3.3b more particles are collected than actually exist in the stream. Thus, $V_n < V_s$ amd $C_m > C_t$.

(3) In Fig. 3.3c fewer particles are collected than actually exist in the stream. Thus, $V_n < V_s$ and $C_m > C_t$.

The effects of improper sampling flowrate for aerosol systems have been studied by a number of workers (refs. 42-45). Improper gas flow will result in an inaccurate particle-size distribution. The control of gas flow for aerosol systems is not critical when sampling particules less than 3-5μm in diameter; smaller particles exhibit less inertial effect and thus cause minimal error in the sample process (ref. 46). Control is also not as stringent when sampling from stagnant or low velocity air masses (e.g., ambient air sampling). However, wind velocity and gravitational effects can interfere with sampling of aerosols from ambient air.

Millipore filters are available which can be used to collect aerosol particulates (ref. 47). They are made of mixed esters of cellulose and have a pore size of 0.8 μm (Type AA). A number of contaminants listed by OSHA (Occupational Safety and Health Administration) can be collected and quantified by gas chromatography (ref. 48). Among these are alpha naphthyl thiourea (ANTU), Toxaphene, chlorinated diphenyl oxide, chlorodiphenyl, DDT, dibutyl phthalate, di-2-ethylhexyl phthalate (DOP), hexachloronaphthalene, octachloronaphthalene, organophosphorus pesticides, pentachlorophenol, terphenyls, and tributylphthalate. Krzymien (ref. 49) has demonstrated the collection of fenitrothion in vapors and aerosols on plates of cascade impactors, and its subsequent determination by gas chromatography using a flame photometric detector. He was able to quantify 10^{-10} g/ℓ in an air sample.

3.2.3 Gases and vapors

The sampling of air is generally more complex than the sampling of liquids. The main reason is that the sample itself is not visible (in most cases) and is quite likely inhomogeneous. Sampling of a plant stack (Section 3.3.3) usually is done within the stack or directly over the top. This is quite different compared to sampling in slowly moving air (e.g., testing pollution in the region of a highway or on plant grounds). In the former (stack sampling), the sample moves past the sampling point but one can be reasonably assured that mixing takes place. In the latter case (slow-moving air), a single sample is unlikely to represent a correct picture of the pollution problem. The sampling point and the time interval should be varied (i.e., multiple samples) before the analytical results have significant meaning.

The person collecting the samples must know the volume of the sample container and/or the flow rate of the sampling system. Which parameter he chooses depends upon the method chosen for sampling. There are a number of procedures for the calibration of flow control devices. The American Society for Testing and Materials Method D-1071-55 gives a good description of most of these techniques (ref. 50). An excellent coverage of flow and volume measurements has been presented by Nelson (ref. 51).

Atmospheric gas sampling requires attention so that an unmodified sample is obtained. Gaseous components in the atmosphere are not subject to the same effects as discussed for aerosols (inertial and electrostatic effects). Stagnant or slow moving air mass (density differential between the gas and air) may result in an uneven distribution of the gas within the air mass. This may be overcome by sampling from a turbulent flow, which may be induced at the sampling site.

Leakage is one of the biggest problems with gas samples, especially stored samples. Gases may diffuse through such materials as rubber, neoprene and plasticized poly(vinyl chloride) (PVC) (refs. 52-53) and containers or seals made of these materials should be evaluated where samples will be stored for periods in excess of a few hours. Some gaseous components also react photochemically; opaque containers for sampling and storage will eliminate this problem.

3.2.3.1 Flow measurements. Air samples can be taken either by the grab technique (see Section 3.3) or by the integrated technique. Integrated sampling is usually part of a preconcentration technique where large volumes are sampled (Sections 3.4 and 3.7). A properly integrated sample requires that the volume of the sample be accurately known, i.e., the sampling rate is constant. A flow meter between the sampler and the pump is the easiest and most frequently used device for this technique. Its greatest disadvantage is that it must be carefully watched and adjusted, since changes in back pressure can alter the flow.

There are a number of ways to maintain a uniform sampling rate with this type of system: (1) Place a restricting orifice on the discharge side of the pump (ref. 54). (2) Sense the air flow rate and alter the pump speed for adjustment (ref. 55). (3) Use an aspirator or siphon bottle; this system must maintain a constant head of pressure to maintain constant sampling rate (ref. 56). (4) Critical orifices may be used to maintain constant flow rate sampling. Commonly used critical orifices are the hypodermic needle (refs. 57 and 58) or glass capillaries (ref. 59). (5) Use a float in conjunction with an orifice (refs. 60 and 61) to maintain constant flow. (6) Use a differential flow controller (ref. 62). It has the advantage of only requiring a 180-torr pressure drop for good operation; critical orifices need approximately 0.5 atm. (7) Use flow meters connected to an adjustable valve (ref. 63 and 64). This equipment is expensive but it makes excellent control possible.

3.2.3.2 Volume measurements. When one obtains a sample of air with a rigid sampling system and then uses the entire volume of the sample for analysis, the container volume must be accurately known. The volume of the container may be obtained by weighing the container empty, then filling it completely with a liquid of known density. One reweighs the container and calculates the volume from the weight and density of the liquid.

An alternative technique is to evacuate the container to a very low known pressure. One connects the container to a known volume of air and attaches a pressure gauge. The known volume of air is allowed to expand into the evacuated container, and the resulting pressure recorded. The volume of the evacuated container can be determined from the initial and final pressure and the volumes of the air-filled container and guage.

3.2.4 Liquids

The main concern when sampling pressurized or public water systems is obtaining a sample which has not been standing quiescent in the system. This quiescent sample may have dissolved materials from exposure to valve packing, check valve, pump fittings and other lubricants. Opening the system for a short period of time will circumvent this problem. The sampling device (container) is generally not critical but it must be clean. The cover for the container should be lined with metal foil (aluminum) or PTFE (polytetrafluoroethylene) to avoid leaching of organic material from the cover.

When samples above ambient temperature are collected, one may use a cooling coil to adjust the sample to approximately ambient temperature. Some test methods require temperature adjustments, and these condition variations should be carried out where indicated.

The exact point of sampling a steam generator depends on the steam generator design, location of chemical and feed lines and other local conditions. A special sampling nozzle is the most desirable method for taking a sample; however, a properly located continuous blow down line may serve as a satisfactory substitute. Locate nozzles or sampling tubes at points remote from confining surfaces and in a submerged position. In selecting the location for nozzles and sampling lines, avoid the possibility of unseparated steam inclusion, excessive amounts of particulate matter, incoming feed water and added chemicals. Significant variation in the composition of matter may exist throughout the steam generator system and, for this reason, samples should be taken from several locations simultaneously. A cooling coil may be necessary to collect these samples.

In the absence of any sampling connections, samples taken from below the water level are not usually representative of the average water and should not be used in the case of controversy. When sampling streams that are under pressure, regulate the rate of flow in the sample line to not less than 500 ml/min after first flushing the sample line sufficiently to remove all sediment and gas pockets.

There are three main procedures for sampling water in the environment: grab sampling, composite sampling and continuous sampling. The general requirements of sampling liquids are the same as mentioned in sampling of gases and vapors. The goal of sampling is to obtain for analysis a truly representative portion of the main body of water. The critical factors necessary to achieve this are points of sampling, frequency of sampling, time of sampling and maintenance of integrity of the sample prior to analysis.

Homogeneity can never be assured so multi-point sampling is usually necessary. A single, most representative sampling point may not be practical to understand the interrelationships in the bulk sample; so a minimum number of sampling points will have to be chosen. Most single-point samples collected from a system should be recognized as unrepresentative to some degree.

3.2.5 Sampling proportional to time and flow

Composite sampling is the combining of individual (grab) samples taken at frequent intervals or by means of automatic samplers and is adaptable to subsequent chemical and physical analysis. One should indicate whether the volume of sample is proportional to the rate of flow or to time. Determinations on the composite sample will represent the average for the stable constituents; variations of unstable constituents should be determined on individual samples. Collect composite samples from process waters during one 24-h period or during at least one complete process cycle if the process is cyclic in nature. This information should be on the sample label. One should choose a suitable flowrate to give the proper volume (ca. 4ℓ) for the composite sample. When samples are taken on a time basis (e.g., from a stream) composite samples usually are composed of equal quantities of daily samples for a suitable number of consecutive days.

When considering whether to sample proportional to time or to flow, one must consider whether the concentration of the pollutant and flow velocity of the stream are steady with time.

If both or one of the units is steady with time, constant sample rate will yield a true average concentration per sampling time. If the concentration and flow velocity are unstable, the sample rate is proportional to flow velocity variations at point of sampling (ref. 65).

Factors which may affect the minimum sample duration time would include: (a) estimated conentration of component(s) to be encountered, (b) rate at which the sampling may be performed, and (c) the sensitivity of the analytical procedure to be used in the determination. As was pointed out before, the end use of the data is a very important criterion and the sampling time should be adjusted accordingly.

3.2.6 Solids

The sampling of solids for environmental purposes should fall into two main categories: the sampling of solids suspended in water or gases and the sampling of solids, e.g., soil samples. The sampling of suspended solids in liquids is mentioned in Section 3.3.2. In this type of system the solids may be filtered from the representative sample and treated as an individual solid sample. In the sampling of solids it is necessary to be concerned about having enough sample to perform the analysis. A number of guidelines are available in the literature (refs. 66 and 67). If the solids are being sampled from a moving stream, the opening of the sample container should be at least three times the size of the largest particles. If the solids to be sampled are not moving (at rest), the minimum weight in kg should be 0.05 times the largest particle (measured in mm). The statistically correct number of samples can be calculated using eqns. 3.5 and 3.6 in Sections 3.2.1.

When collecting solids which are suspended in a gas, the filter should precede the gas and/or absorption containers. Care must be taken so that the first collector (i.e., the filter) does not interfere with any sample collectors which follow. Use of a synthetic atmosphere will assist one in determining the effectiveness of the sampling system. To obtain accurate data from such a procedure the collection and measurement steps must be performed under controlled conditions.

In general, a large number of locations for the collection of intermittent samples will provide more reliable data than samples collected on a continuous basis at a small number of sampling sites.

A number of techniques are available to obtain solid samples from environmental sources:

(1) Allowing the particles to settle because of gravity. This technique is widely used in air pollution studies (refs. 68 and 69).

(2) Filtering out the solids. This may be used for either air or water samples. The filtering material should be chosen so as not to react with the sample particles (ref. 70).

(3) Passing the particules through a narrow bore orifice and then through a solution. Collection rates may vary from 0.75 to 1.5 ft^3/min. Collection efficiency is almost 100% (refs. 71 and 72). This procedure is predominantly used in air pollution studies.

(4) A high electrical difference is maintained across two electrodes (12,000 to 45,000 V d.c.). Particles collected by the electrostatic precipitation technique are easily removed from the electrodes (washing or brushing techniques) (ref. 73). A thermal gradient may also be used for the collection of the particles. In thermal precipitation the small particles migrate to the lower temperature zone (ref. 74).

Sampling devices which remove particulate material from the air are never completely efficient (i.e., 100%). The reason is that all sizes of particles may be present; thus, one should choose a collection technique which is efficient for the particle size range of interest.

3.2.6.1 Soils. A number of sampling methods are available for soil samples. We will consider three of the most common: auger-boring sampling, split-barrel sampling and thin-walled tube sampling.

Auger-boring sampling requires that the auger penetrate the desired distance into the soil. The soil is removed for examination and analysis. The process is repeated until enough sample is obtained. The soil structure may be destroyed and the moisture content changed by this type sampling. All samples should be sealed in an appropriate container and labeled clearly. If more than one type of soil is sampled with the auger, be sure to clean the sampler each time and prepare a separate container for each type of soil.

This sampling method is limited only by the soil characteristics and ground-water conditions. It has the advantage of being simple, being used for any purpose where disturbed samples may be taken and being usable for ground-level determination and indication of changes in strata. The disadvantages are that it is easy to contaminate the sample and that does not provide a reproducible sampling unit with uniform cross-section to the desired depth.

In split-barrel sampling of soils it is necessary to clear out a hole. When sampling saturated sands or silts, withdraw the drill bit slowly. This prevents loosening the soil around the hole. A minimum depth of 18 in. (0.45 m) is sufficient penetration into the soil. The sampling operation should be repeated at intervals of 5 ft. (1.5m) in homogeneous strata and at every point where a change of strata is noted. The sampler is brought to the surface, opened and a careful description of the soil sample is recorded (i.e., composition, structure, consistency, color and condition). Samples are then put into jars, sealed (with wax or hermetically sealed) to prevent evaporation of the soil moisture. The soil samples should be protected against extreme temperature changes.

This sampling method can not be used for soils containing rock. It does provide representative samples of soils for laboratory testing, gives a measure of the resistance of the soil to penetration of the sampler, provides an uncontaminated sample and can be used on hard surface soils. The disadvantages are that one needs equipment to clear out the hole before using the sampler and if sampling on the hard soil surface, precautions must be taken to insure that the falling weight (depth of penetration per blow of hammer) is not reduced by friction.

The same precautions as noted above for split-barrel sampling should be followed for thin-walled tube sampling. In this technique, the sampling tube is pushed into the soil by a continuous and rapid motion (no impacting or twisting). When the soils are so hard that a direct pushing motion will not penetrate, a slight driving action with a light-weight hammer is permitted (or one should use the split-barrel sampling technique). Before withdrawing the tube, turn it at least two revolutions to shear off the bottom. The samplig operation is performed exactly as described above for the split-barrel sampling technique.

The thin-walled sampling tube cannot be used with stony, dry, or sandy soils. It has the advantages that it can be used to obtain relatively undisturbed soil samples which are suitable for laboratory testing, provides uncontaminated samples, provides reproducible sampling units with approximately uniform cross-section to the desired depth and does provide a representative sample. As with the split-barrel sampling, one needs the equipment to clear out the hole before sampling.

3.2.6.2 Particulates. Air-metering devices are required to relate the quantity of pollutant collected to the total volume of gas sample. Thus, a precise and accurate measurement of the volume of gas drawn through the sampling device is needed. Flow measurement is made downstream from the collecting device, i.e., between the collecting device and the vacuum source. This arrangement makes it necessary to measure the pressure and temperature to calculate the correct volume. No correction for gas leakage from the pump is required if flow is measured into the pump. There are two types of gas measurement devices: (a) ratemeters, e.g., orifices, nozzles, venturimeters, and rotameters; and (b) volume (positive displacement) meters, e.g., wet test meters, dry test meters, and cycloid gas meters. (See Sections 3.2.3.1 and 3.2.3.2 for flow and volume measurements). Instantaneous volume flowrate is measured by ratemeters. They have the disadvantage of frequent monitoring to be sure an accurate calculation of the total volume is made. Their advantage is that they require little space. Integrated total volume passing through the system is measured by volume meters. Thus, they provide a direct record of air volume sampled. The dry test meter is most commonly used in this type of sampling because it is economical, light-weight and fairly sturdy. These meters are accurate to ±1.0%.

Collection devices for particulate air samples require measurement of the quantity (volume) of air passing through the system. Vacuum sources are commonly used, e.g., vacuum pumps. There are two major types of vacuum pumps: positive displacement and centrifugal. Positive displacement pumps are linear with respect to suction pressure (head) and capacity, whereas the centrifugal types are non-linear. Collection of particulates in flowing gases is carried out as depicted in Fig. 3.3 (Section 3.2.2).

The high volume sampler method may be used in three modes:

Rotameter mode

The high-volume sampler is fitted with a clean filter and run for at least 5 min. A rotameter is attached and the ball adjusted to read 65. The adjusting mechanism is then sealed so that it can not be changed. Shut off the motor, remove the filter and attach an orifice calibration unit (see Fig. 3.4 for diagram of a typical orifice meter) in its place. Operate the high-volume sampler at a series of different, but constant, air flows (minimum of 6). Record the differential manometer reading on the orifice calibration unit and also record the reading of the rotameter at each air flow. Convert the differential manometer reading of the orifice calibration unit to m^3/min. The flow rate of an orifice meter may be calculated by use of Poiseuille's law (refs. 75 and 76).

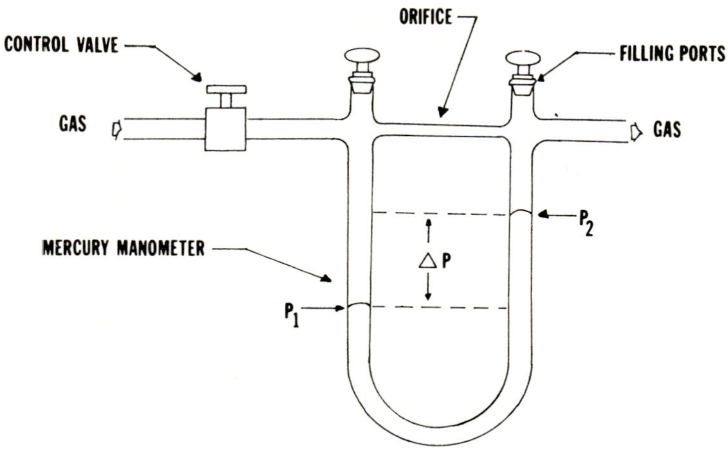

Fig. 3.4 Drawing of an orifice meter. Reprinted by permission of the publisher (Ann Arbor Science Publishers Inc.)

$$Q = \frac{(P_1 - P_2)(1 + (P_1 - P_2)/2P_2}{r} \tag{3.14}$$

where Q = flow rate (ml/sec). To convert to m^3/min, multiply by (60 sec/min) (1 $m^3/10^6$ ml);

P_1 = upstream pressure (dynes/cm^2);

P_2 = downstream pressure (dynes/cm^2);

r = capillary flow resistance (g/cm^4-sec) = $(128\eta L))/\pi d^4$;

L = capillary length (cm);

d = internal diameter capillary (cm);

η = viscosity of gas (poise or g/cm-sec); viscosity of air at 18°C=182.7 micropoises. To convert micropoises to poises,multiply by 10^{-6};

π = 3.1416.

Plot rotameter reading versus Q. Identify this plot with the high-volume sampler and rotameter serial number for future use (the orifice must remain clean and otherwise unaltered).

Draft guage mode

The high volume sampler is fitted with an orifice calibration unit. Run for at least 5 min, then attach the draft guage to the fitting provided in the multi-holed orifice plate built into the discharge of the sampler (this is the fitting used for the rotameter). Operate the high-volume sampler at a series of different, but constant, air flows (minimum of 6). Record the differential manometer readings on the orifice calibration unit and that on the draft guage at each air flow. Convert the differential manometer readings to m^3/min, Q, using Eqn. 3.14. Plot draft-guage readings versus Q. Identify the plot with the high-volume sampler and draft-gauge serial numbers for future use.

Recorder mode

The high-volume sampler is connected exactly as in the two previously described modes, except a recorder is installed in place of the rotameter or draft-gauge. Run the unit for at least 5 min and operate the high-volume sampler at a series of different, but constant, air flows (minimum of 6). Record the differential manometer readings on the orifice calibration unit and that of the recorder at each air flow. Convert the differential manometer readings to m^3/min, Q, using Eqn. 3.14. Plot the recorder readings versus Q. Identify this plot with the high-volume sampler and recorder serial numbers for future use.

If the pressure or temperature during the calibration of the high-volume sampler is substantially different from the pressure or temperature during the orifice calibration, a correction of the flow rate, Q, may be required. If necessary, the corrected flow rate may be calculated from the following equation:

$$Q_2 = Q_1(T_2P_1/T_1P_2)^{1/2} \quad (3.15)$$

where Q_2 = the corrected flow rate in m^3/min;
Q_1 = the flow rate during the high-volume sampler calibration (m^3/min);
T_1 = the absolute temperature during orifice unit calibration (°K);
T_2 = the absolute temperature during high-volume sampler calibration (°K);
P_1 = the barometric pressure orifice unit calibration (Torr);
P_2 = the barometric pressure during high-volume sampler calibration (Torr).

This correction applies only to orifice meters having a constant orifice coefficient.

3.3 GRAB SAMPLING

When one does not wish an elaborate sampling program or a field monitoring program, the grab sampling technique is used. This sampling technique may be carried out for air samples with the use of evacuated bulbs or inflatable plastic bags. Large-volume hypodermic syringes are commonly employed to obtain grab samples from the atmosphere. These may be used dry or have an absorbing solution in the barrel. For water sampling, an appropriate size rigid container may be used. The volume of most containers is determined by calculation or by a liquid displacement technique. Halogenated hydrocarbons (fluorinated and chlorinated) in the air were collected by this technique in metal containers and analyzed in an extremely clean stainless steel sampling and calibration system (ref. 77).

3.3.1 Air

Flexible containers. Plastic bags such as aluminized Scotch-pak, Mylar® and Saran are the typical materials used. The sample may enter the bag by use of Tygon or glass tubing or even an ordinary tire valve. Decay curves should be run on the substance sampled so a reasonable storage time can be established. The sample is introduced into the bag by a hand-operated pump or a squeeze bulb. Bags can be reused after purging with clean air and checking for any residual components. Certain contaminants can neither be sampled nor stored in any type plastic bag because of their reactivity. Reactivity of a substance should be determined in the laboratory prior to field use. If storage is necessary, the bags should be protected from direct sunlight.

A review of much of the work done using bag sampling has been published by

Schuette (ref. 76), in which he addressed the factors which are important for maintaining a representative sample (i.e., integrity of the sample). The important factors are the concentration of the gas, the surface area of the bag, relative humidity, temperature, physical and chemical properties of the bag material, as well as the presence of reactive components in the sample. Many of the factors listed can also be important for other types of containers as well, e.g., glass, metal, etc. However, if appropriate precautions are taken, one can be assured of having a representative sample. This is borne out by the number of investigators who have used this type of container very successfully (refs. 78-84). Smith and Pierce did a study of the application of plastic bags for benzene, methanol, methyl isobutyl ketone and methylene chloride at concentrations which bracketed the 1969 TLV (Threshhold Limit Values) (ref. 83).

Rigid containers. Evacuated glass bulbs may be prepared by use of a vacuum pump and the neck sealed by heating and drawing to a tip (see Fig. 3.5). These are suitable for sampling atmospheres for analysis of gases such as carbon dioxide, oxygen, methane, carbon monoxide, hydrogen and nitrogen. They are not suitable for sampling very reactive gases such as hydrogen sulfide, oxides of nitrogen or sulfur because such gases may react with dust particles, moisture, the wax sealing compound, and even with glass of the container. By the time the sample is analyzed, the proportions of these gases will have been altered.

A similar evacuated container may be used to collect samples directly in a chemical absorbent. The bulb differs from the conventional type illustrated in Fig. 3.5 in that it contains a liquid absorbent for the particular gas to be analyzed. The liquid absorbent dissolves and preserves the gas sample in a form that may be determined by chemical analysis.

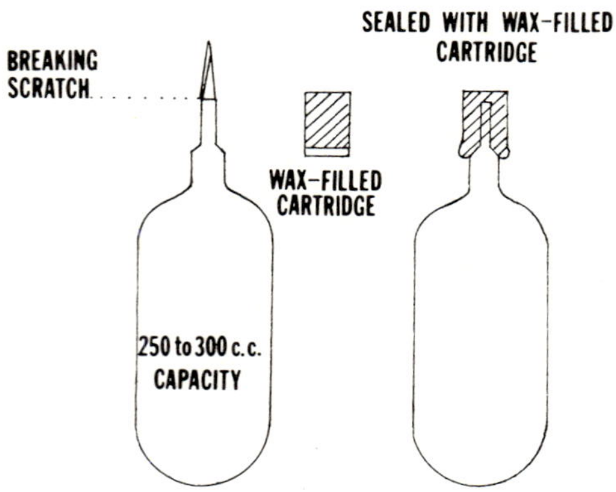

Fig. 3.5 Rigid evacuated sample container.

To prepare a sample bulb such as shown in Fig. 3.5, the bulb should be evacuated to a pressure of 1 Torr or less. If it is not possible to reach such a low pressure, the bulb should be prepared in an uncontaminated atmosphere and the evacuated pressure obtained should be measured with a mercury manometer. Once evacuated, the flask is sealed at the tip with a torch and removed from the vacuum system. The volume of sample, at barometric pressure may be calculated as follows:

$$V_s = V_b(P_1 - P_2)/P_1 \qquad (3.16)$$

where V_s = volume of sample at barometric pressure;
V_b = volume of the bulb;
P_1 = barometric pressure;
P_2 = pressure in evacuated bulb.

To collect a sample of gas with this type of container, one breaks the tip at the scratch mark provided on the neck of the container. The surrounding atmosphere immediately enters and fills it to atmospheric pressure. The container is then sealed by forcing a wax-filled cartridge over the open neck.

Two additional sampling set-ups to obtain grab samples are depicted in Fig. 3.6. In Fig. 3.6a, an evacuated bulb is used to obtain the sample. If using this particular system, the sample volume may be determined by use of the ideal gas law after recording the temperature and pressure before and after the sampling is completed. The following equation may be used for this calculation.

$$V_s = (V_f - V_r)(T_s/P_s)(P_i/T_i) - (P_f/T_f) \qquad (3.17)$$

where V_s = volume of gas at STP;
V_f = volume of flask;
V_r = volume of absorbing solution, if any;
T_s = temperature at standard conditions, 0° C(273° K);
P_s = pressure at standard conditions;
P_i = initial pressure in flask (Torr);
P_f = final pressure in flask (Torr);
T_i = initial temperature in flask (°K);
T_f = final temperature in flask (°K).

Fig. 3.6b shows the use of a flexible bag for the grab sample. When using this technique a vacuum is drawn on the rigid sample box which in turn causes the flexible bag to inflate, thus drawing in a grab sample of the atmosphere. Once the sample has been collected, the bag is sealed and transported back to the laboratory for analysis. The rigid sample container serves a two-fold purpose: it is the

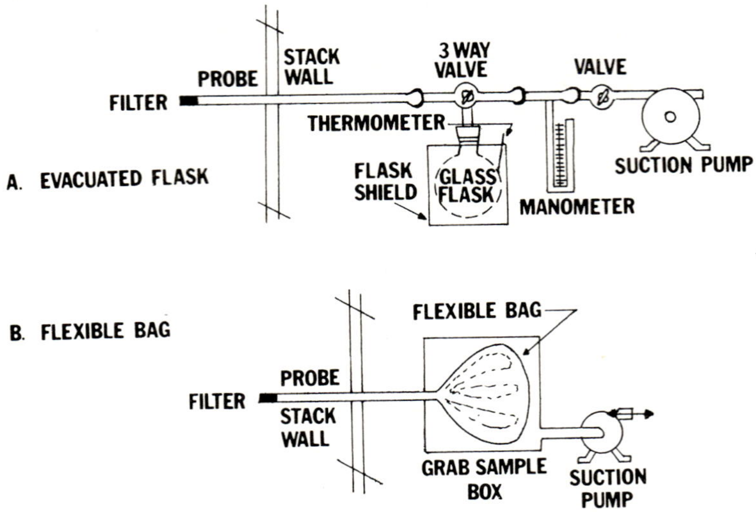

Fig. 3.6 Systems for grab sampling; (a) evacuated bulb, (b) flexible bag. Reprinted from Air Pollution Control, John Wiley & Sons, Inc.

container during the sampling with a flexible sampling bag and a protective covering for the bag and sample for the transport to the laboratory. A modification of this type sampling would be to surround the flexible bag with water. During the sampling process, the water is allowed to drain at a steady rate.

Gas samples may also be collected by displacement (Fig. 3.7). The gas sampler may be metal or glass, closed with stopcocks or screw clamps and rubber tubing if the sample does not react with rubber. One may pinch the clamps and rubber tubing if the sample does not react with rubber. Pinch clamps are not satisfactory, as they may not have enough tension to close the rubber tubing completely. Metal containers should be constructed of inert metal, e.g., stainless steel. Iron or tinned-iron sample containers are prone to rusting and consume too much oxygen.

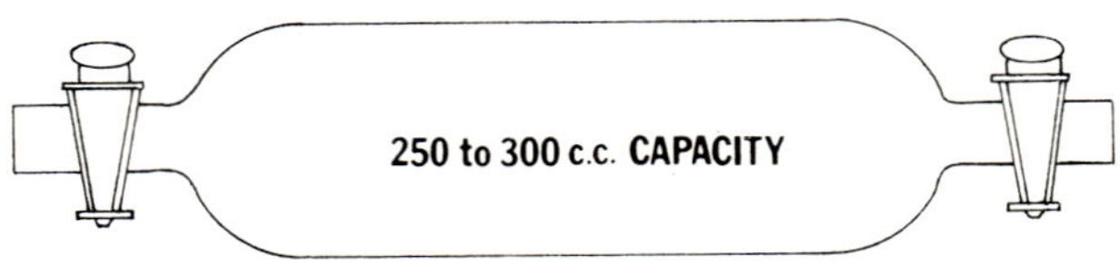

Fig. 3.7 Displacement sample tube.

Metal containers are not suitable for sampling of many substances (e.g., H_2S, SO_2, or oxides of nitrogen) because the gases react with the metal and are lost from the sample. Copper is reactive with acetylene and acts as a catalyst for a number of gas-phase reactions. To fill any of the above containers it is necessary for the original air or gas content to be completely swept out and replaced by the gas to be sampled. This may be accomplished by pulling the gas sample through for a period of time or completely evacuating the tube. An aspirating device such as a double-acting rubber bulb aspirator or a double-acting foot pump may be used. An aspirator may also be made of a bottle or can filled with water, as shown in Fig. 3.8.

To sample by gas displacement, a double-acting rubber bulb may be connected so that pressure and relief on the rubber bulb will draw the original content of air out of the sample container and admit the atmospheric sample (see Fig. 3.9). The compression-aspiration process should be repeated about thirty-five times to replace the air in a 250 ml container. Before using this technique, check the performance by compressing the bulb and noting whether there is any leakage.

This sampling technique is limited to those contaminants for which sensitive analytical methods are available or the contaminant in the test atmosphere is at high concentration. The technique has the advantage that the volume of air sampled need not be accurately known, requires relatively simple equipment, offers variety and flexibility to many sampling applications and is inexpensive.

Some disadvantages of this technique are the small sample size, sample loss on walls of the container, analysis for reactive gases must be performed quickly, leaks in sample container to exterior and contamination by leaks from outside to interior of container. Since the sample is not concentrated, the technique is more sensitive to contamination by leaks.

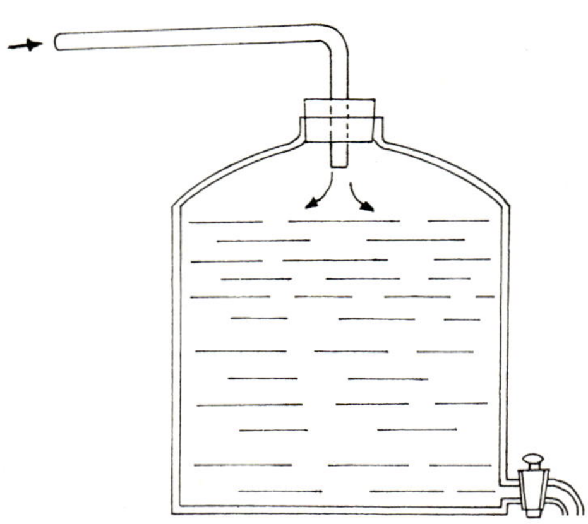

Fig. 3.8 Water-filled bottle (aspirator) for collection of gas sample by displacement.

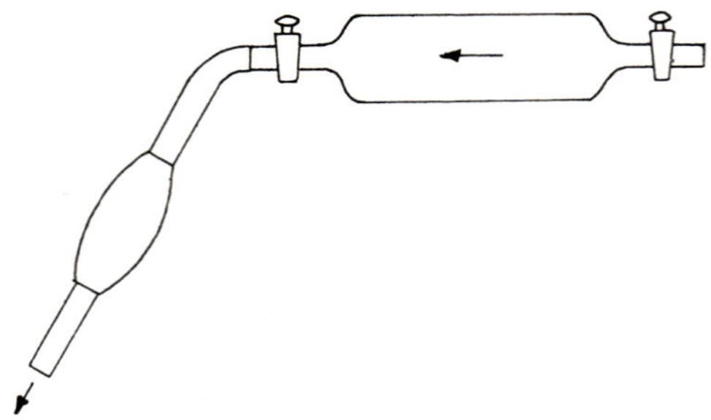

Fig. 3.9 Double-acting rubber bulb to admit gas into sample container.

3.3.2 Liquids

Grab sampling is easily applied to water from wells, rivers, streams, lakes, pipelines, processing tanks, steam generators, etc. at atmospheric pressure or higher pressures for chemical analysis. A grab sample represents only the conditions existing at the point and time of the sampling. The sample container should be rinsed several times with the water to be sampled before taking the sample.

The point of sampling should be chosen with extreme care; avoid surface scum. A wide variety of conditions is found in large bodies of water so it is not possible to recommend the exact point of sampling. Ordinarily, samples are taken at a number of points and combined to obtain an integrated (composite) sample for such large bodies of water. Allow sufficient distance downstream, with respect to stream flow at the time of sampling, from a tributary or source of pollution to permit thorough mixing. If this is not possible, sample the stream above the source of pollution and a distance of 1 to 3 miles below the source.

The composition of raw water from a large body of water may be determined using individual samples at random intervals. Samples taken from the shoreline of a lake or river should be taken at shorter time intervals.

Water undergoing treatment must be sampled at a frequency adequate for control. The sampling time is directly related to the rate at which critical characteristics can become difficult to control.

Normally water samples are taken without separation of particulate matter and thus the presence of colloidal or flocculent material should be included so that they are present in representative proportion. One should collect a sample volume of 2 to 4ℓ. When sampling highly radioactive water, a smaller sample may be desirable to reduce the radiation hazard.

Samples should be collected at least 0.5 mile below dams or waterfalls to allow for the escape of entrained air. Unless one is interested in the effects of local conditions (e.g., inlet streams, stagnant areas) be sure to avoid these unrepresentative areas when sampling.

If air contact would cause a change in the concentration or characteristics of a constituent to be determined, sample without air contact and completely fill the container. Sampling of unconfined water at specific depths (e.g., ponds, lagoons, reservoirs, etc.) during which air contact or agitation would cause a change in the characteristics of a constituent should be carried out with an apparatus in which the sample flows through a tube to the bottom of the container. A volume equal to four to ten times the volume of the sampling container should pass through before the sample is taken. If dissolved gases are of no interest, any less complicated apparatus that will permit the collection of the sample at a desired depth or a composite sample from several points in a vertical section will be adequate.

When samples are to be shipped, do not fill the bottle completely if a possibility of freezing may occur. An air space of approximately 25ml usually suffices for this purpose. If volatiles are to be determined, the sample should fill the container, which then should be sealed tightly and protected from freezing.

If the water to be sampled is at low pressure, special means must be provided to extract the sample. Installation of a barometric leg is the simplest arrangement. A positive-displacement type pump can be arranged to discharge into the sampling container at atmospheric pressure. The sample receiver may be placed between the pump and the sampling point. An arrangement of this type permits one to equalize the receiver to atmospheric pressure and drain into the sampling container.

Grab sampling is a simple and inexpensive technique suitable for bacteriological analysis. A limitation is that grab samples represent the conditions existing only at the point and time of sampling.

3.3.2.1 Composite sampling. As stated in Section 3.2.5, composite sampling is the combining of individual grab samples. This type of water sampling is not suited for bacteriological assays and is limited only to short-lived radionucleides. It is a simple and inexpensive sampling technique suitable for chemical and physical analyses when one is interested in average sample data., i.e., data presenting an average value for the concentration of various constituents. It is not a good sampling technique for the determination of unstable or volatile components of the water sample (e.g., carbon dioxide, chlorine, oxygen, etc.).

3.3.2.2 Continuous sampling. This procedure may be employed to sample water from sources such as wells, reservoirs, pipelines, processing tanks and steam generators on a continuous basis for chemical, physical and/or radiological analyses.

When simultaneous samples from several locations in the same water source are required, the water may be drawn continually from each point (proportional to flow) and mixed into a single sample. Particulate matter often accounts for some of the variables (e.g., size, quantity and type) which introduce errors in the analysis, especially if they are disturbed. The water system should flow fast enough to maintain the heavier particles in suspension and the volume should be large enough to prevent undesirable filter action. Pumps employing suction principles disturb the gas-liquid balance of dissolved gases such as oxygen or carbon dioxide and these types should be replaced by submersible pumping types.

The time response of a system (i.e., the time required for the system to reach 63.2% of the total change between the initial state of equilibrium and the final state of equilibrium resulting from a step change introduced at the input to the system) describes the sampling system's ability to respond to transient conditions in the water source.

This sampling procedure has essentially no limitations. It is suitable for on-stream analyzers, chemical, physical and radiological analyses. Particulate matter causes little error and the technique possesses a high degree of accuracy. The main drawback is that the equipment is more complicated and more expensive than other grab sampling techniques.

3.3.3 Stack sampling

Stack sampling is usually done to monitor a source of suspected pollution or where it is desired to measure the efficiency of collection equipment. No matter what the purpose of the study, the sampling procedure must be capable of a reasonably accurate evaluation of the source expressed in terms of the rate of emission of specific pollutants. Stack sampling may be necessary: (1) to obtain data about the emissions for an emission source or to identify a predominant source in the area (e.g., to determine the hydrocarbon release from an organic solvent used in a degreasing tank), (2) to determine compliance with regulations, (3) to gather information for the proper selection of control equipment, (4) to determine efficiency of control equipment installed to reduce emissions (e.g., if a device is guaranteed to be 95% efficient for removal of particulate matter the effluent stream must be sampled at the inlet and outlet of the device to determine if the guarantee has been met), (5) to evaluate emission changes resulting from process or equipment modifications, (6) to collect data for legal evidence.

The laboratory sample size dictates the sampling procedure to be employed. Stack gases may be collected by absorption, adsorption, freeze-out, and grab sampling. Particulate matter sampling is more difficult due to particle size, charge and shape. A satisfactory sample would be one that is representative of the flow pattern in the duct and stack, is free of extraneous contamination and shows no evidence of chemical or physical change. In stack sampling, one measures the rate of air or gas flow through the duct, stack, roof monitor , window or other cross-section at which pollution concentrations are to be sampled. If particulates are to be sampled, the approximate range of stack or duct velocity must be known prior to sampling. This enables one to use the isokinetic sampling procedures. A number of instruments for measuring air velocity are available (ref. 86-89); the standard pitot tube (ref. 90) combination with a differential pressure gauge is the most widely used and the most satisfactory velocity-measuring instrument.

Obtaining a representative sample from a stack is a complex task because all process parameters vary with both time and location within the stack. The conditions in a stack may be described by three related terms: (1) pollutant concentration, (2) stack gas flow rate, (3) pollutant mass emission rate. The pollutant mass emission term is the most important because many regulations specify permissible emissions on this basis. The three terms may be related as:

$$\overline{PMR}_s = \bar{C}_s \times \bar{F}_s \qquad (3.18)$$

where $\overline{PMR}_s$ = the average pollutant mass emission rate;
$\bar{C}_s$ = the average stack pollutant concentration;
$\bar{F}_s$ = the average rate of volumetric stack gas flow.

The average rate of volumetric stack gas flow, $\bar{F}_s$ is determined by measuring the average gas velocity, $\bar{V}_s$, and the area of the stack, A_s.

$$\bar{F}_s = \bar{V}_s \times A_s \qquad (3.19)$$

Basically stack measurements may be divided into two categories: those internal and those external to the stack. The internal measurements are temperature, pressure and flow rate of the exhaust gas. External measurements are gas sample temperature, flow rate and total flow. These are measured to establish the proper sampling rate and determine the total quantity of sample withdrawn.

A good discussion of stack sampling may be found in Volume III of Stern's multi-volume treatise on Air Pollution (ref. 91). Archinger and Shiegehara (ref. 63) characterized sources according to their variations in cross-section and time. Stack pollutant concentration can have time and cross-sectional variation because $\bar{C}_s$ and $\bar{F}_s$ must be measured in such a way that the resulting average pollutant mass emission rate, $\overline{PMR}_s$ is representative (i.e., unbiased).

When source conditions are steady (i.e., no variation with time) and uniform (i.e., no variation from point to point within the cross section), only one measurement is needed for accurate results. Sampling for a gaseous pollutant falls into this category.

When source conditions are steady with regard to time and non-uniform with regard to velocity, a series of measurements must be made. The number of measurements is determined by the shape and dimensions of the stack and its proximity to various obstructions such as elbows. A coal-fired power plant measured for particulate emissions is an example of this category. Usually a composite sample is extracted. A composite sample is one on which a series of pollutant measurements is made on the sample collected in the same pollutant collection device, as opposed to using a separate collector for each measurement.

When source conditions are unsteady with time and uniform with velocity, measurements are needed at only one location within the cross section. The longer the sampling period for noncyclic operation the more representative the measurement will be (e.g., sampling of a cyclic operation for a gaseous pollutant at a point where the gas stream is not quiescent). If the source (time) and the flow conditions (velocity) are both non-uniform, one must use more complicated sampling procedures. Under these conditions it is very difficult to obtain a representative sample.

The cross-sectional concentration of gaseous pollutants in a stack is usually uniform because of the diffusion and turbulent mixing processes which are present. Thus, with the concentration conditions uniform, it is necessary to sample at only one point in the stack to determine the average concentration, $\bar{C}_s$. The gas samples may be taken on an integrated sample flow rate basis or a grab sample basis to meet the sampling strategy requirements decided upon before the project was undertaken. The basic components used in these sampling trains are a filter, a probe, gaseous pollutant collector, gas meter, flow regulator and a gas mover. Thermometers and manometers may also be used to monitor the temperature and pressure. Tne filter removes the particulate matter that may interfere in the gaseous component analysis (see Fig. 3.10). The gas meter measures the sampling flow rate and/or the total gas volume sampled, while the gas mover (pump or other vacuum source) provides the suction to move the gas sample through the other sampling components.

Bias occurs if isokinetic sampling is not achieved; sampling rates for particulates are above or below the proportional rate. A particulate sampling train (Fig. 3.11) is composed of a nozzle, probe, a device for collecting the particulate matter, one or more gas meters and a gas mover. Generally a pitot tube is used to measure the stack gas velocity (see Section 3.2.3.1). A number of operational principles may be used for the collector, i.e., filtration, electrical precipitation, liquid scrubbing, condensation or any combination of these.

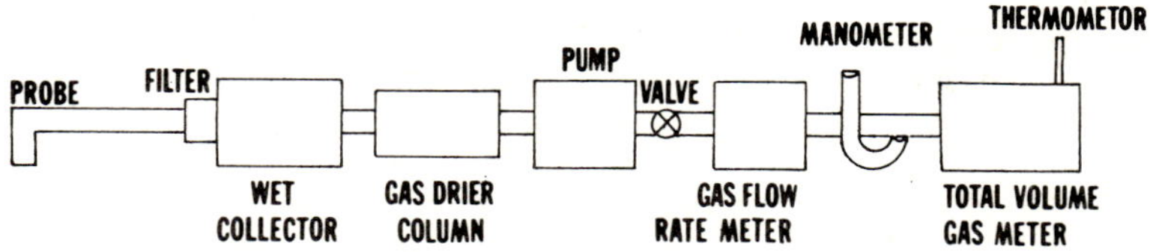

Fig. 3.10 Absorption set-up for gaseous pollutants in a stack. Reprinted by permission of the publisher (Ann Arbor Science Publishers Inc.).

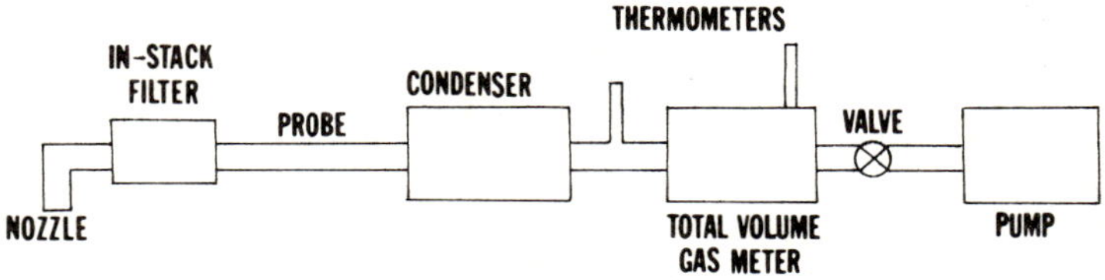

Fig. 3.11 A particulate sampling train. Reprinted by permission of the publisher (Ann Arbor Science Publishers Inc.).

If no regulation exists for the sampling strategy, then one must be selected. A preliminary survey would determine the source category. Discussions of this type are beyond the scope of this book. Information concerning such surveys may be found in Chapters 3 and 5 of the book by Brenchley et al. (ref. 88).

3.4 ADSORPTION TECHNIQUES

The use of adsorbents is a unique technique to isolate a variety of pollutant gases. The technique is not selective for specific gases because the commonly used adsorbents have not been developed to that level. Consequently, a mixture of gases is usually retained which may cause difficulties when the sample is analyzed in the laboratory. Commonly used adsorbents are: activated alumina, silica gel, molecular sieves, porous polymers, ion-exchange resins, charcoal and Celite®. The samples are drawn through a container and over the adsorbent maintained at either ambient or subamient temperatures. The adsorbed materials can be removed by heating the adsorbent to volatilize the trapped material (then flushed with an inert gas into a chromatograph or other analysis system) or by removal from the adsorbent with a suitable solvent. Analysis may be made in a number of different ways.

A number of problems are associated with the technique. Water vapor can deactivate the adsorbent or the adsorbent can cause isomerization or reaction to occur and thus alter the sample (refs. 79, 89 and 90). Because of the nonspecificity of the adsorbent concentrated gases on the surface can react with each other or because of the instability of certain gases and adsorbents decomposition can occur when sorbates are removed.

Adsorption sampling has been used as a first step in the measurement of a wide variety of gases, e.g., organic gases (ref. 91), radioiodine (ref. 92), volcanic gases (ref. 93), nitrous oxide (ref. 94), ammonia and amines (ref. 95), carbon monoxide (ref. 96), carbon dioxide (ref. 97), ozone (ref. 98) and the pesticide lindane (ref. 99). Versino et al. (ref. 100) described the sampling of organic air pollutants by what they refer to as "dynamic enrichment" on an adsorption column. They used this technique because it was judged by others to be reliable. The basis for reliability is that (1) the adsorbed components can be thermally eluted (avoids the need for ultra-pure solvents), (2) the adsorption and desorption procedures are easily automated. (3) the technique allows complete adsorption of components that have a retention volume lower than the sample volume passing through the adsorption column and an equilibrium concentration proportional to the concentration in the original sample for more volatile components, and (4) the sampling can be performed in situ. The porous polymer, Tenax GC (poly-p-2,6-diphenylphenyleneoxide, surface area ca. 30 m^2/g, 60/80 mesh) was used as the adsorbent because it is hydrophobic, has low adsorption strength, allows the thermal elution of relatively high boiling compounds, has good thermal stability and does not react with most organic compounds.

An excellent review of the technique of extracting organic compounds from water by means of the adsorption sampling technique has been presented by Dressler (ref. 101).

3.4.1 Gases

Gases and vapors will adhere to some extent to any solid surface at ordinary or low temperatures. Porous solids not only have exterior surfaces but interior surfaces as well, and some such solids possess vast networks of extremely minute channels and submicroscopic pores. These solids include activated carbon (non-polar), silica gel (polar) and activated alumina (polar). Each differs in the number and kinds of substances they adsorb, as well as in the amount of sorbed substances they will retain. Selection of an adsorbent requires consideration of a number of properties of the adsorbent. It should have a high specific surface area. Alumina (Al_2O_3) and silica gel (SiO_2) usually range from 200-600 m^2/g, whereas activated carbon can vary from 500-2000 m^2/g, depending on the activation process. The adsorbent should have some selectivity towards polar and non-polar sorbates; e.g., alumina and silica gel have great affinity for polar compounds whereas charcoal (carbon) has more affinity for non-polar compounds. There should be no chemical reactivity between sorbents and sorbates unles one desires this property to enhance selectivity. Retention capacity for the sorbates should be high and the sorbent should not easily fracture or crush. On the other hand, the desorption rate should allow rapid quantitative recovery of the sample.

Silica gel and activated alumina may be used for short duration sampling from atmospheres that are sufficiently dry so that the adsorbent does not become saturated with moisture. Activated carbon is capable of adsorbing organic gases and vapors in preference to atmospheric moisture. The carbon has a greater over-all capacity for adsorbing and retaining atmospheric gases and vapors but silica gel exhibits a greater selectivity. Activated carbon is more suitable for collecting a mixture of gases and vapors, especially when present in small concentrations. Desorption of gases and vapors from silica gel is easier than from activated carbon, and it is possible to obtain some separation of gases and vapors by selective desorption from silica gel. Passing contaminant-free air at temperatures up to 350°C has been found to be effective in removing gases and vapors sorbed on silica gel. Other effective methods include extraction with polar solvents such as water or alcohols and distillation with saturated steam at ambient pressures. Each adsorbent can be suitably surface modified by other salts to make it suitable for chemisorption (refs. 102 and 103) of gases and vapors.

Adsorption using activated carbon follows the general scheme that gases having critical temperatures below -50°C and boiling points below -150°C are essentially not adsorbed at ordinary temperatures. Gases and vapors with critical temperatures between 0°C and 120°C and boiling points between -100°C and 0°C are moderately adsorbed, while heavier gases and vapors (boiling points above 0°C) are readily adsorbed and retained at ordinary temperatures (e.g., odorous organic and inorganic substances).

One of the simplest isolation and concentration techniques, applicable to air and gaseous samples, consists of a trap immersed in a cold bath to condense the components of interest. This type of trap (see Fig. 3.12) is more effective if it is packed with some solid material, thus presenting a large surface on which condensation may occur. Materials that may be used for this purpose are alumina, silica gel, charcoal, porous polymers and even liquid-coated chromatographic support materials (e.g., coated diatomaceous earths).

Most of the chromatographic supply companies sell prepackaged and presealed adsorption tubes which may be used for this type of sampling (SKC, Inc., Eighty Four, PA, U.S.A. and Supelco, Bellefonte, PA, U.S.A. are two). The adsorption tube is taken to the laboratory where it may either be connected directly to the injection port of the gas chromatograph and the sorbed gases removed while carrier gas is flowing through the tube or the gases may be desorbed and condensed in a cold trap to be injected as a liquid. Proper desorption techniques will permit the use of the tube for subsequent sampling. Fig. 3.13 is one possible arrangement which may be

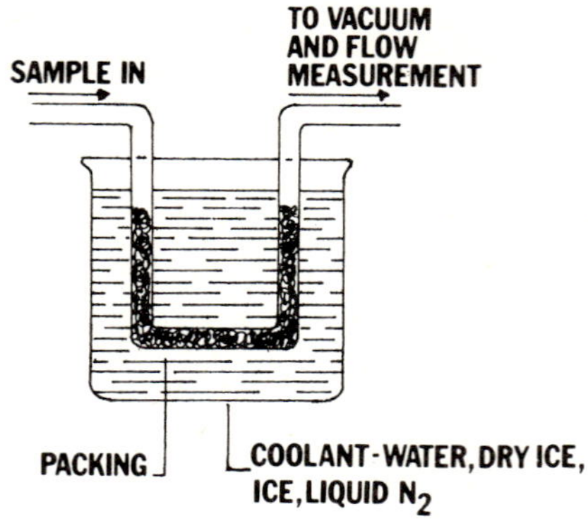

Fig. 3.12 Tube trap to condense and concentrate gas samples. Reprinted from Modern Practice of Gas Chromatography, John Wiley & Sons, Inc.

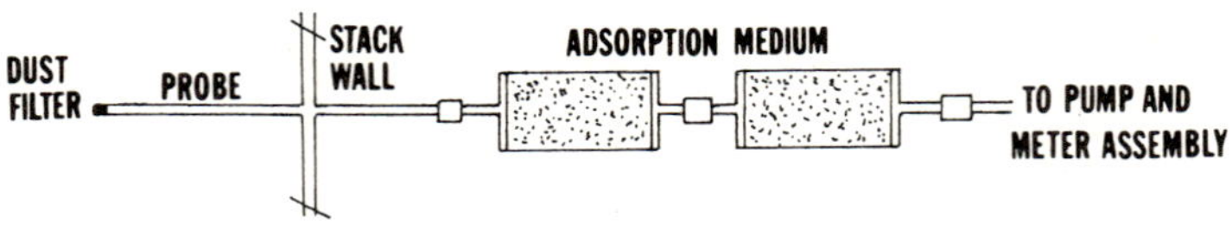

Fig. 3.13 Gas adsorption system. Reprinted from Air Pollution Control, John Wiley & Sons, Inc.

used for this type of sampling.

The adsorbed gases and vapors may be desorbed by displacement with superheated steam and/or heating the adsorbent under vacuum and distilling the desorbed material into cold traps. The extent of adsorption is inversely related to temperature and quantity of gas adsorbed. The adsorbed gases may also be removed by use of the non-organic solvent carbon disulfide and the sample analyzed as a solution. Some limitations of this technique are that it is not always possible to remove the adsorbed gas without decomposition and a mixture of gases is usually retained which can cause analytical difficulties. Adsorption is a simple, inexpensive technique capable of isolation of a desired component from other constituents and concentrating trace amounts of a sample component. The adsorbent often can be easily deactivated by water vapor in the sample stream. Certain adsorbents are known to cause isomerizatio and/or hydrolysis and thus alter the sample whereas removal of some adsorbates cause chemical changes, e.g., esters hydrolyze when steam is the eluent.

Table 3.2 is a representative listing of the many types of adsorbents which have been used and the types of compounds which have been sampled.

TABLE 3.2

Sorbents used to samples various components in air and/or enclosed atmospheres

DMSO = dimethyl-sulfoxide; ECD = electron-capture detector;
FID = flame ionization detector; FPD = flame photometric detector;
GC = gas chromatography; HPLC = high-performance liquid chromatography;
MS = mass spectrometry; RSD = relative standard deviation;
TCE = trichloroethylene. In ppb and ppt, the American Billion (10^9) and trillion (10^{12}) are meant.

Sorbent	Sorbates	Comments	Ref.
ALUMINA	Halogenated alkanes, n-alkanes, alcohols, nitrogen-containing compounds	Thermal stripping, FID	104,105
	Benezene in air	Thermal desorption, N_2 carrier, FID, Detection limit: 0.05 mg/m^3;RSD=2.5%	106,107
	Glutaraldehyde	Water-ultrasonic desorption, FID.	108,109
CARBON Charcoal	Hydrocarbons	Charcoal coated on ferro-magnetic wire. Results compared to porous polymers and support bonded phases.	110,111
	Aromatic hydrocarbons	Use of Bendix Flasher for desportion	112,113
	Benzene	CS_2 solvent removal, back-flush technique. RSD = 0.62%	114,115
Ambersorb XE-340	$CHCl_3$,C_5H_{12}, CH_3OH	FID, hot-wire detector. Back flush technique with He.	116
Carbosieve B	Halogenated n-alkanes, vinyl chloride, dimethyl ether	Thermal desorption ppb sensitivity. FID and ECD.	117
Activated carbon	Halogenated alkanes, n-alkanes, alcohols, nitrogen-containing	Thermal stripping FID	104,105

TABLE 3.2 (Cont'd.)

	compounds.		
	Alkanes, alkenes, alkynes	0.2-1000ppb. Thermal desorption, FID	118
Granular Active Carbon	Organic vapors	Pyrolysis wire coated with active carbon. Thermal desorption.	119
Activated Carbon	Benzene, toluene	Personal air sampling devices. GC-MS. Solvent desorption. Detection limits: 1.0 ppm.	120
Dried Active Carbon	Aromatic hydrocarbons	CS_2 desorption N_2 carrier. FID.	121
Activated Carbon	s-Tetrachloroethane & other chlorinated hydrocarbons.	Thermal desorption	122
Packed Charcoal	Vinyl chloride	CS_2 desorption. N_2 carrier. FID	123
Carbopack BHT	Low-molecular-weight organics.	Thermal elution, FID	124
Ambersorb XE-340	Low-molecular-weight organics	Thermal elution, FID	125
Carbochrome	C_1-C_{13} organic compounds. 136 compounds checked.	Thermal desorption	126
Carbosieve B/ Porapak QS	Volatiles from chicken manure	GC-MS	127
LIQUID SUBSTRATES			
Dexsil 300GC/ Chromosorb	Organic pollutants	Thermogradient desorption GC-MS	80
DC-200 Silicone oil	Chlorpyrifos and Ronnel (both phosphorus compounds).	Thermal gradient. ppb sensitivity. FPD	117
Dinonyl phthalate on Parolith	Halothane	Desorbed at 100°C. Ar carrier, FID	128
C_{22} Firebrick polyethylene glycol, 400/ Chromosorb W.	Vinyl chloride	CH_3OH-CO_2 coolant. Thermal desorption. N_2 carrier. FID	129
OV-101/Chromosorb, Porasil D	C_5-C_{12} Hydrocarbons Sulfur compounds	Analysis on poly(phenylether)+H_3PO_4.	130

TABLE 3.2 (Cont'd.)

PMS-100/ Silica gel	H_2S	Thermal desorption	131
PFMS-4/Celite 545	Dimethyl terephthalate.	Thermal desorption. Temp.programmed, 55-135°C.	132
GC Durapak-Carbowax 400/ Porasil F	Chlorpyrifos + pesticides	$(C_2H_5)_2O$ extraction. ECD.ppt conc.	133
OV-17	Volatiles from triolein heated in air	82 peaks obtained. FID and MS.	134,135
MOLECULAR SIEVES			
Molecular Sieve 13X	Acrolein	3-8μ g/g(97 ± 11% recovery)60-200μg/g (90 ± 7% recovery)Dynamic dilution of samples.FID.	134,135
Molecular Sieve 5A	SO_2,H_2S	Thermal desorption.FPD. Results compared to Porapak P, Q; Carbopack B, Tenax, Chromosorb 102, Molecular sieve 13X.	136
POLYMERS			
Porous Polymers	Halogenated hydrocarbons	Essentially 100% recovery. Flow: 0.05-2.0 ℓ/min.	137
Chromosorb 106	$CHCl_3$,C_5H_{12},CH_3OH	FID,hot-wire detectors. Backflush technique with He.	116
Porapak N and R	Aromatic hydrocarbons, halogenated n-alkanes phenols, alcohols,aldehydes, ketones.	Thermal desorption ppb sensitivity.FID and FPD	117
Porapak P and Q	Halogenated alkanes, n-alkanes, alcohols nitrogen-containing organic compounds.	Thermal stripping.FID	104
	Sulfur compounds, aliphatic animes, carbonyl compounds, hydrocarbons, alcohols, phenols, fatty acids, indoles.	ppt to ppb range	138
Chromosorb 102	Pesticides	Extraction with acetone-hexane, ECD.Efficiency of sorption and desorption 70-96%	139
Porapak Q	Halogenated hydrocarbons	Thermal desorption	140

TABLE 3.2 (Cont'd.)

Chromosorbs 101,102,103	Halogenated hydrocarbons	Thermal desorption	140
Porous Polyurethane Foam (PPF)	Pesticides vapors	Solvent extraction with toluene.ECD	141
Porapak Q	Halocarbons	Elution extraction.ECD	142
Porapak N	CH_4,C_2H_4,NO	Headspace over soil	143
Porapak Q	Combustion and pyrolysis products from polymers	Collection done at 2 temps: CO_2-acetone(H_2S,HCN,C_4 hydrocarbons),-196°C (carbon oxides,alkanes).	144
SILICA GEL	Halogenated alkanes, n-alkanes,alcohols, nitrogen-containing compounds.	FID.Thermal desorption	104,105
	Organic solvent vapors: acetone,methanol,TCE, benzene,toluene.	Aqueous DMSO desorbent >90% recovery. RSD=4.9 - 8.7%	145
	Nitrotoluene and toluidine isomers	CH_3OH elution,N_2 carrier,FID	146
	Aniline compounds	C_2H_5OH extraction. Heptanol internal std. He carrier.FID	147
	Volcanic gases, H_2S, SO_2	Thermal desorption, 130°C	148
	4,4'Methylenebis (2-chloroaniline)	CH_3OH extraction. 10µl injected into HPLC. Reversed-phase system.UV detection.Quant. 0.15 µg/sample.Precision: 9.2% at 1.5 µg/sample and 14% at 0.15 µg/sample.	149
SUPPORT-BONDED CHROMOSORBS	High-molecular-weight organic volatiles	Sorbates removed by solvent extraction	150
TENAX POLY-P-2,6-DIPHENYLPHENYLENEOXIDE)GC	Acetone, diphenylamine petroleum,n-C_6	Recoveries from 90-100%.RSD=5% at ppb level.	151,152
	Aromatic hydrocarbons	Use of Bendix Flasher desorption.	112,113
	$CHCl_3$,C_5H_{12},CH_3OH	FID, hot wire detectors.Backflush technique with He.	116

TABLE 3.2 (Cont'd.)

	Aromatic hydrocarbons, halogenated alkanes, phenols.	Thermal desorption. ppb sensitivity.FID and PFD.	117
	Halogenated alkanes, n-alkanes,alcohols nitrogen-containing compounds.	Thermal stripping. FID.	104
	Phenols and cresols	94-98%collection efficiency.ppb.FID.	152
	Food volatiles from rice and soybeans	Thermal desorption	153
	Odorous compounds from axilla area of humans. i-propyl esters of fatty acids, aldehydes, anti-oxidants.	GC-MS. Thermal desorption	154
	Alcohols,ketones, acids,S-,NO_2-,Cl-, NH_2-compounds, phenols.	Solvent extraction. Liquid injection for phenols.Others thermal desorption.	155
	Cigarette smoke volatiles:isoprene, acetaldehyde, acrolein.	Thermal desorption	156
	Insect chemosensory compounds		157
	Halogenated hydrocarbons	Thermal desorption	140
	Vinyl chloride	Thermal desorption N_2 carrier.FID	158
	Low-molecular-weight organics	Thermal elution. FID.	125
	Esters,aldehydes, ketones,alcohols, acids,anhydrides, amines,hydrocarbons, chlorinated hydrocarbons	Calculated breakthrough volumes and safe sampling volumes.	159
	Sulfur dioxide	Sampling rate: 20-50 ml/min. polyphenylether+ H_3PO_4 on 40/60 Chromosorb T, 2m x 3.2 mm column, recovery >95%.Limit of detection, 2 ppb, ambient temp. trapping	160
XAD-2,	Halogenated alkanes,	Thermal stripping	104,105

TABLE 3.2 (Cont'd.)

XAD-4 XAD-7 Resins	n-alkanes,alcohols, nitrogen-containing organic compounds	FID.	
XAD-4 Resin	Halogenated alkanes, alkenes	C_2H_5 elution 60-95% recoveries. ECD.	161
XAD-4 Resin	Organophosphorus pesticides	Alkali FID. Sample efficiency ca. 84±4%.	162
MISCELLANEOUS			
Potassium nitrate	Dichlorvos	FPD. Confined air locations. Recoveries - 80%.	163
Gold-coated glass beads	H_2S,dimethyl sulfide methyl sulfide,R-S compounds	Thermal desorption FPD.limits of detection:0.1 ppt.	164
Polished copper mesh	Chlorinated hydrocarbons	Parts per trillion. Collection at -196°C. Analysis on Porapak Q.MS detection.	165
Nylon gauze	Malathion	Acetone-hexane (1:1) extraction FPD.	166

3.4.2 Water

Adsorption sampling techniques are much more efficient for gases than for liquids. Liquids may be sampled more easily by grab sampling (Section 3.3.2), dipper techniques (Section 3.8), etc. The adsorption technique, for water sampling, has been used directly and indirectly. In the direct mode, the water to be sampled is passed through a column or porous polymer or similar adsorbent where the components of interest are removed from the water stream. The volume of water passed over the adsorbent should be monitored so that component(s) concentraton(s) can be related to sample volume. In the indirect mode, the water sample is stripped of its volatile components by bubbling an inert gas (N_2 or He) through the water sample. The gas stream is then passed through a column containing charcoal, a porous polymer, alumina or silica gel. Knowing the volume of water which has been sparged one can calculate the concentration of the volatile components after analysis. Whether the water is sampled directly or indirectly by the adsorption technique, the subsequent removal of sorbates is achieved by heating the adsorbent, steam stripping or solvent extraction (leaching). The gas-stripping technique has been used by a number of workers in the field (refs. 99 and 102). A closed loop technique of gas-stripping and removal of organics (CAM, i.e., carbon adsorption method) has been used by Grob and Grob (ref.

151). The direct adsorption technique has been used by many workers (refs. 106, 108, 110, and 112).

The volume of water required as a sample will depend upon the levels of contamination. Usually 500-1500 ml is adequate. The sampling rate for direct adsorption should be approximately 5-7 ml/min (depending on sorbent-bed depth and cross-sectional area). Fig. 3.14 represents one possible arrangement for the sampling, desorption, separation and determination scheme of a water sample of organic material. The limitations and disadvantages are discussed in Section 3.4.1 apply to water samplings well as air sampling.

Table 3.3 lists many of the adsorbents which have been used and the types of components which have been sampled by this technique.

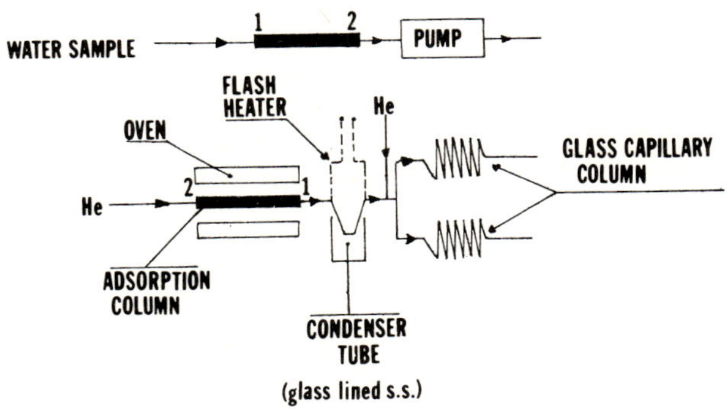

Fig. 3.14 Adsorption and elution system for water and gas samples. Reprinted by permission from J. Chromatogr., 122(1976) 373-388.

3.5 CHEMICAL REACTION TECHNIQUES

A sampling technique which involves a chemical reaction is used when the component to be determined does not exist in a form that is readily detected by a chromatographic method. Many such techniques are available where the final measuring step is a spectroscopic one (ref. 178). There are the Kitagawa gas detector tubes, the Drager analyzer tubes, the Gastec detector tubes, the NBS detector tubes and the Specific chemical detector tubes.

TABLE 3.3

Sorbents used to sample various components in water samples

See Table 3.2 for abbreviations

Sorbent	Sorbates	Comments	Ref.
CARBON			
Ambersorb XE-340	$CHCl_3$, C_5H_{12}, CH_3OH	Thermal desorption with He backflush,FID, hot-wire detectors.	116
Spherocarb	Organic contaminants	Organic solvent elution, waste and drinking waters	167
MOLECULAR SIEVES			
13X	Nitrous oxide	Adsorption at 0°C, thermal desorption. Molecular sieve 5A for determination.EDC	168
POROUS POLYMERS			
Chromosorb 106	$CHCl_3$,C_5H_{12},CH_3OH	Thermal desorption with He blackflush.FID, hot-wire detectors.	116
Macroporous polymer	Aromatic hydrocarbons	ppb levels. Thermal desorption.	169
Porapak Q and Chromosorb 101	Vinyl chloride	FID	170
Styrene-divinylbenzene	Pesticides		171
Porous Polyurethane Foam (PPF).	Pesticides	DC-200/Gas Chromatogr Q 160°C. ECD.	172
TENAX-GC	$CHCl_3$,C_5H_{12},CH_3OH	Thermal desorption with He backflush.FID,hot-wire detectors.	116
	Petroleum, n-C_6, acetone, diphenylamine	Recoveries varied from 90-100%.RSD=5% at ppb level	151
	Haloforms	Thermal desorption.GS-MS ppb levels.	172
	Haloforms	Thermal desorption, FID,ECD.	173

TABLE 3.3 (Cont'd.)

	Volatile organic compounds from marine biota	n-C_5 elution. SE-30 SCOT column. GS-MS. Poly nuclear aromatic hydrocarbons done by LC.	174
	Ozonization products from waste water. C_6-, C_7-, C_8-, C_9- aldehydes	Nitrogen sparged samples	175
XAD-RESINS			
XAD-2	Haloforms	Pyridine elutions.ECD	176
XAD-4	Organic contaminants	Organic solvent elution. Waste and drinking water samples.	167
XAD-2	2,3,7,8-Tetrachlorobenzo-p-dioxin (TCDD)	10^{-11} to 10^{-12} g/ℓ C_2H_5OH elution. GC-MS. Recoveries of 52-84% with 0.0001-0.15 μg standards.	177

If a gas will react chemically with a liquid, adsorption efficiency tends to be much higher than would be expected from purely physical adsorption (ref.179). Ethylene has been sampled by the adsorption technique using charcoal which has been impregnated with bromine. This converts the ethylene to ethylene dibromide which is more easily retained (ref. 180). Adsorption on activated charcoal may be increased by the impregnation of other chemicals on its surface; e.g., alkaline pollutants are more readily sorbed when the charcoal has been impregnated with sulfuric or phosphoric acid. Hydrogen sulfide can be readily sorbed and concentrated if the charcoal has been impregnated with lead acetate. The last two examples do not leave the sorbed components readily available for gas chromatographic analysis. They would, however, be more available for ion-exchange chromatography.

In chemisorption the rate is very high due to the high efficiency of active surfaces as compared to small areas. As a consequence, smaller amounts of pollutants can be trapped. This results in a smaller air volume being required or a shorter sampling time. Another result is that low concentrations of toxic substances may be determined by the sampling of large air volumes.

Traces of water in a gas sample may be determined by the use of a flame ionization detector if the gas sample is collected by passing over calcium carbide. The product from the reaction of water and calcium carbide is acetylene, a substance readily detected by flame ionization.

3.6 FREEZE-OUT OR CRYOGENIC TECHNIQUES

The collection and concentration of atmospheric gases and vapors by condensation at low temperatures has some advantages over other concentration methods. Collected material is immediately available for further separation or analysis, without requiring removal of solvents or desorption from an adsorbent. Condensation is generally the most reliable method for preserving sampled gases and vapors (i.e., no chemical reaction with any part of the collecting device or intra and intermolecular reactions).

The main disadvantage of collection by condensation is the large quantities of water that are usually condensed with the collected vapors (see Fig. 3.13). However, condensation of atmospheric contaminants with water is not generally a serious problem because their extraction from water is experimentally straightfoward. A special problem in low temperature collection techniques is caused by the formation of condensation mists from the cooled air. Such mists are composed of solid or liquid particles (aerosols) and often pass through the cold traps in sufficient quantities to reduce the collection efficiency of the equipment. A simple filter, such as a glass wool plug, in the cold trap minimizes such losses. As sample collection proceeds, the solid condensate acts as a filter and collection efficiency increases until the resistance of the equipment to flow becomes too great.

Selection priorities of refrigerants for condensation or freeze-out techniques are given to those materials which maintain a constant temperature by change of phase. Some of the more common refrigerants available are given in Table 3.4. In addition to the listing in Table 3.4 other intermediate temperatures are available by use of slushes (ref. 181). A slush may be prepared by placing the solvent in a Dewar flask and adding liquid nitrogen with stirring until the system turns to a slush (see Table 3.5). Liquid nitrogen refrigerant may condense atmospheric oxygen in the trap, especially at low sampling rates. The condensed oxygen may then react with a number of the condensed components. One selects a refrigerant sufficiently cold that the vapor pressure of any trapped material will be low enough to prevent significant evaporation during the sampling. Generally, the vapor pressure of condensed gases and vapors should be about 1 Torr (133 Pa) or lower at trap temperature. Air sometimes is trapped with the condensed sample which can make the sample less stable. Flushing the trap with helium usually removes the trapped air but also will flush out some of the sample components (ref. 182). The technique most frequently employed uses a series of traps at progressively lower temperatures. Thus, some degree of fractional condensation can be achieved. If the last trap is empty, one has some assurance that the sampling was efficient. Shepperd et al. (ref. 183) described an assembly for trapping contaminants in air. The collection assembly used a drying tube following the sample trap. The outlet of the drying tube led to a condenser which could be cooled in liquid nitrogen. By controlling the temperature of the sample trap one could do a fractional distillation of the sample components into the condenser. After distillation at a given temperature is complete, the fraction in the condenser is evaporated directly into the inlet system of a gas chromatograph or mass spectrometer. Problems associated with this procedure include the adsorption of acidic and unsaturated organic vapors by the Ascarite in the drying tube and the collection of some water in the condenser.

This technique is limited to samples having low partial pressures, e.g. >2 Torr (266 Pā). Partial pressures below 1 Torr (133 Pa) prevent any material from being available for further separation or analysis, without requiring the removal of solvents from the adsorbent. It may be used for a wide range of organics as well as at both ambient and source emission monitoring conditions.

Schmidt et al. (ref. 186) used this technique to sample very volatile substances from a pressurized pilot plant. The sample was bled off by means of a needle valve and then condensed in a methanol-carbon dioxide cooled evacuated container. After collection the vacuum is relieved and the sample system is allowed to reach thermal equilibrium. Deimel and Dulson (ref. 187) condensed air samples taken along highways and then separated and analyzed the condensate by gas chromatography. They were able to identify 38 hydrocarbons.

Angerer et al. (ref. 188) collected air samples in coiled stainless steel which was immersed in liquid nitrogen. Gas chromatography-mass spectrometry was the technique of detection for the components in air. This type of sampling ("pressure application") allows the separation to be carried out at a lower temperature. Chloromethane has been frozen out of air streams by use of a loop packed with glass

TABLE 3.4

Common refrigerants for use in condensation sampling techniques

Refrigerant system equilibrium	Temperature (°C)
N_2(liquid) ⇌ N_2(gas)	-195.0
O_2(liquid) ⇌ O_2(gas)*	-183.0
Air(liquid) ⇌ Air(gas)	-147.0
CS_2(solid) ⇌ CS_2(liquid)	-118.5
Dry Ice(CO_2(s))-Acetone	-80.0
CO_2(liquid) ⇌ CO_2(gas)	-78.5
NH_3(liquid) ⇌ NH_3(gas)	-33.4
Ice water-salt	-16.0
H_2O(solid) ⇌ H_2O(liquid)	0

*requires careful handling

beads and immersed in liquid oxygen (ref. 189). Vaporization into a column of Durapak n-octane on Porasil, separation of sample components and monitoring of the effluent by mass spectrometry at m/e 50 completed the analysis. Trace odorants (sulfur compounds, lower aliphatic amines, carbonyl compounds, hydrocarbons, aliphatic mono-alcohols, phenols, lower fatty acids and indoles) at ppt-ppb levels in air, were trapped using liquid oxygen or trapping on porous polymer beads at ambient temperature (ref. 190). The air sample volume varied from 1-5/ℓ. Detection limits of the technique were 0.05-2 ppb.

TABLE 3.5

Cold bath slushes (ref. 181)

Slush	Temperature (°C)
Carbon tetrachloride slush	-23.0
Chlorobenzene slush	-45.0
Chloroform slush	-64.0
Ethyl acetate slush	-84.0
Toluene slush	-95.0
Methyl cyclohexane slush	-126.0
n-Pentane slush	-130.0
Isopentane slush	-160.0

3.7 TRAPPING TECHNIQUES

Trapping techniques are taken to include bubble traps which are fitted with fritted discs, bubblers and/or diffusers. To operate such sampling equipment it is necessary that some sort of pumping device be available to move the sample through the trap. Appropriate gas pumps should have adequate capacity and perform uniformly. Mechanical pumps (electric) are used for prolonged periods of sampling operation. Induction motors are preferred since they operate more uniformly even with variations in line load.

Hand pumps (large glass syringes or piston pumps) are suitable for sampling small volumes or when gas flow need not be constant. Aspirators are usable where there is an adequate supply either of water or air under constant pressure. Siphons may be used when small sample volumes are adequate or when very low sampling rates are needed. It is desirable that the gas inlet tube of the siphon bottle extend to the bottom to maintain a uniform flow as the liquid level falls.

Gas metering devices are required when the sample is pulled by means of an aspirator or a pump that acts to maintain a reduced pressure downstream of the sampler. These meters may either be volume meters or rate meters.

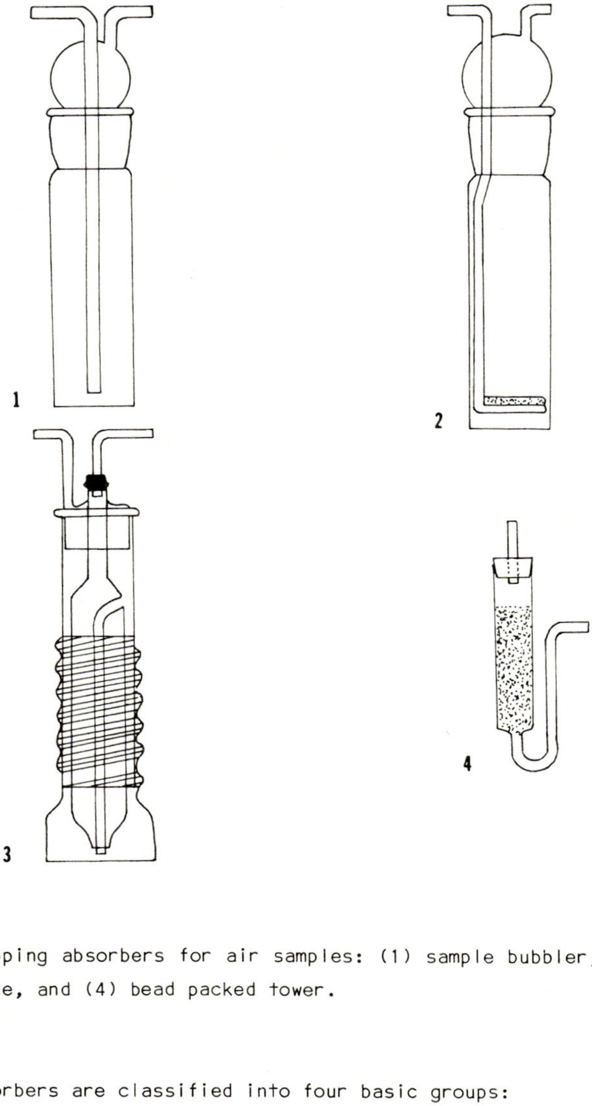

Fig. 3.15 Trapping absorbers for air samples: (1) sample bubbler, (2) diffuser type, (3) spiral type, and (4) bead packed tower.

Absorbers are classified into four basic groups:

(1) Simple bubbler having an absorbent capacity of 5 to 100 ml and sampling flow-rates of 5 to 3000 ml/min. They are very simple in design, usually non-plugging but have a short gas-liquid contact time.

(2) Diffuser bubblers having an absorbent capacity of 1 to 100 ml and sampling flow-rates of 500 to 100,000 ml/min. They are easy to use and have good gas-liquid contact time, but are subject to plugging.

(3) Spiral absorbers having an absorbent capacity of 10 to 100 ml and sampling flow-rates of 40 to 500 ml/min. They are fairly efficient at low flow-rates.

(4) Bead-packed towers have an absorbent capacity of 5 to 50 ml and sampling flow rates of 500 to 2000 ml/min. Their flow rates are variable with resistance and are only efficient at lower flow-rates. Each of these absorbers is depicted in Fig. 3.15a-d.

It may be necessary to extend tubing beyond the sampler in order to reach the desired point for sampling. In such cases, the tubing must be large enough not to offer high resistance to gas flow. The tubing must be appropriate for special conditions such as high temperatures and they must not adsorb or absorb appreciable portions of any sample components. No single material is usable for all lines; some metals, glass and varióus plastics have frequently been used. Every material, however, is suspect until it has been demonstrated to have minimal interference.

If air to be sampled is connected through a system with tubing which extracts some part of the sample, it is essential to avoid any recirculation of such sampled air through the extraction equipment. The air that has passed through the equipment must be ejected into a space from which sampling is impossible. In outdoor sampling this is accomplished by simply exhausting the air downwind. In still air, vent the exhaust by a tube or duct or blow it by a fan away from the intake location.

The sampler is always used in series with a gas-metering device which may be followed by a pressure gauge or manometer. A pressure gauge or manometer serves several useful purposes: it enables one to decide whether the gas volume should be corrected because of too low a pressure for acceptable accuracy and in turn furnishes the data for such a correction. The pressure measurement is necessary if there is a likelihood of increased resistance to gas flow in the absorber (from ice formation, precipitation or presence of particules). When the gas being sampled contains particulate matter, an appropriate filter should be placed ahead of the sampler. The filter should precede any metering or gas-collecting equipment and should be nonreactive and nonabsorbing to the gases being sampled. Dry fibrous glass, porous plastic films and cellulosic paper have proven satisfactory. Siliceous material or cellulosic paper should not be used when hydrogen fluoride is being sampled. Non-reactive plastics (e.g., polymers of vinylidene chloride, ethylene or tetrafluoroethylene) may be used to obviate this difficulty. The gas meter and the pump should be downstream from the sampler to prevent loss of the gas constituent being sampled and to minimize contamination of the samples or dilution of the samples by extraneous air.

Collection of particulate matter with the gases and vapors is undesirable because it may interfere with the chemical analysis that follows the sampling or interfere with the collection or metering of the sample.

All meters and sampling assemblies must be calibrated regularly to assure accuracy. The basic apparatus for this calibration is the gasmeter (both volume and rate types).

Absorption of gases or vapors is a solubility phenomenon whereby gas molecules are preferentially dissolved in a liquid-collecting phase. Therefore, this technique should be used only with gases and vapors which are soluble in the absorbing solution. The technique is limited to use with low rate of gas flow. By proper selection of absorbing solution this sampling method can be used to separate the desired constituent from contaminants, concentrate trace components from a sample and offer simplicity and low cost. Problems arise because of increased resistance to gas flow due to any particulates, evaporation rate of the solvent and the small size of the sample that is available.

3.8 DIPPING TECHNIQUES AND TUBE SAMPLING

Dipper sampling is applicable for sampling liquids and semi-liquids where a free or open discharge stream exists, as in small filling and transfer piplines (2 in. in diameter or less) and filling apparatus for barrels, packages and cans. The dipper should have a flared bowl and a handle of a convenient length made of such material (e.g., tinned steel) that will not affect the product being sampled. The dipper must have a capacity adequate to obtain a representative sample and should be protected from dust and dirt when not in use.

The dipper is inserted into the free-flowing stream to collect a portion from the full cross-section of the stream. The gross amount of sample collected should be approximately 0.1%, but not more than 40 gallons, of the total quantity being sampled. The sample portions are poured into a clean, dry sample container as soon as collected. The container must be kept closed, except during a dipper-portion transfer. When all portions of the sample have been collected, the container is closed, labeled and delivered to the laboratory for analysis.

Since the dipper-sampling technique is limited to samples from open streams (when one wishes an instantaneous or grab sample), barrel, packages and cans, samples obtained by this technique cannot be considered as representative samples. Sample volumes are usually small and the technique is not always suitable for taking samples at lower levels of the containers.

Tube sampling is useful for sampling liquids and semi-fluid materials in drums, barrels and cans (these types of containers are becoming more of an environmental problem because of waste-disposal regulations). A glass or metal tube, designed to reach to within 1/4 in. of the bottom with capacity of approximately 0.5 to 1.0 ℓ may be used. To sample a barrel or a drum, stand the container upright and sample from the top through the open bung hole. Containers of 5 gallons (18ℓ) capacity or larger should be sampled in a similar manner using a tube of proportionately smaller dimensions. Containers of less than 5 gallons (18ℓ) capacity should not be sampled but the entire container taken as the sample. If a large number of these smaller containers are to be analyzed, then they are selected at random or in accordance with agreement among the parties involved. An example sampling table is shown in Table 3.6.

The tube sampling technique is limited to use with an easily accessible opening in a container. This is a technique suitable, under proper circumstances, for taking representative samples and very adaptable to securing small samples of liquids.

3.9 HEADSPACE SAMPLING

Headspace sampling is a technique which is more appropriately discussed in the chapter on Sample Treatment (Chapter 4). This sampling procedure is used mainly to obtain samples of the volatile components above a liquid (or solution) and sometimes above a solid. Thus, the sampling procedure is performed in the laboratory as opposed to field sampling. However, if one is going to use the headspace sampling technique there are necessary precautions which must be followed during the sampling of the water supply or solid material.

A number of applications of this technique become evident when one considers the reason(s) for the analyses. A manufacturer concerned with solvent residues in a product: odors from various fruits and foods; alcoholic beverages, tobaccos, pharmaceutical products, gases present in packaged goods (odors from container and/or contents), petroleum products, polymeric substances (unreacted starting materials), physiochemical studies of gas phase equilibria, volatiles in plant effluents, rivers, lakes, wells and sewage and waste products.

TABLE 3.6

Minimum number of container to be selected for sampling from a lot (ref. 191)

Number of containers	Number to be sampled	Number of containers	Number to be sampled
1 to 3	All	1001-1331	11
4 to 64	4	1332-1728	12
65 to 125	5	1729-2197	13
126 to 216	6	2198-2744	14
217 to 343	7	2745-3375	15
344 to 512	8	3376-4096	16
513 to 729	9	4097-4913	17
730 to 1000	10	6860-00	20

Coupling of this technique with gas chromatography makes possible analyses and sensitivity not otherwise available without involved processing of the samples. The outstanding feature of headspace sampling is the absence of a large solvent peak which may completely obscure smaller peaks on the front or tail of the solvent peak. Thus, we will concern ourselves with precautions and/or safeguards during the bulk-sampling step, if headspace sampling (Section 4.3) will be used in the laboratory.

(1) If sampling from a liquid bulk sample (tank storage, stream, well, etc.), use a rigid (glass preferred to metal) container and fill to top. Immediately seal the container with a top which has an inert liner (PTFE or aluminum foil), Label the container and store in an ice cooler to transport to the laboratory.

(2) If sampling a solid material which has small enough particles to pack well, fill container to top, seal and label. Store in a cooler for transport back to the laboratory. If the solid material particles are not small enough to pack well, use a rigid container large enough to hold the entire sample, seal, label and cool. Large-size samples also may be placed in aluminized plastic bags which can be sealed, labeled and cooled.

(3) An alternate sampling procedure for liquids is to add enough salt (NaCl, K_2CO_3, Na_2SO_4, etc.) to the container before going into the field to obtain the bulk sample. The amount of salt added should be enough to form a saturated solution when the container is half-filled with the liquid (especially if sampling water supplies which may have low concentrations (<1 ppm) of volatile components. The container is immediately sealed, labeled and placed in a cooler for transport to the laboratory. The presence of the large concentration of salt increases the ionic strength of the sample and thereby decreases the solubility of the organic components in the sample (e.g., hydrocarbons - both aliphatic and aromatic). NOTE: If this procedure is used during the bulk sampling phase, the top of the container must have an opening (fitted with a septum) to allow for easy entry of the sampling syringe in the laboratory. Do not perform the headspace sampling technique for this sample until it has reached ambient conditions in the laboratory.

REFERENCES

1 H. B. N. Hynes, The Biology of Polluted Waters, Liverpool University Press, Liverpool, 1960.

2 L. Klein (Ed.) River Pollution, II, Causes and Effects, Butterworths, London 1962.

3 K. M. Mackenthun and W. M. Ingram, Biological Associated Problems in Freshwater Environments - Their Identification, Investigation and Control, Federal Water Pollution Control Administration, Washington, DC 1967.

4 C. N. Sawyer and P. L. McCarty, Chemistry for Sanitary Engineers, 2nd ed., McGraw-Hill, New York 1967.

5 L. E. Keup, W. M Ingram and K. M. Mackenthun, Biology of Water Pollution: A Collection of Selected Papers on Stream Pollution, Waste-Water and Water

Treatment, Publ. No. CWA-3, Federal Water Pollution Control Administration, Washington, DC 1967.
6 P. S. Welch, Limnology, 2nd ed., McGraw-Hill, New York 1952.
7 G. E. Hutchinson, A Treatise on Limnology, Vol. 1, Geography, Physics and Chemistry, Wiley, New York, 1957
8 G. E. Hutchinson, A Treatise on Limnology, Vol. 2, Introduction to Lake Biology and the Limnoplankton, Wiley, New York, 1967.
9 F. Ruttner, Fundamentals of Limnology (transl. by D. G. Frey and F. E. J. Fry), University of Toronto Press, Toronto 1953.
10 G. K. Reid, Ecology of Inland Waters and Estuaries, Reinhold, New York, 1961.
11 Standard Methods for the Examination of Water and Wastewater, 12th ed., American Public Health Association, New York 1965.
12 P. S. Welch, Limnological Methods, Blakiston, Philadelphia, PA, 1948.
13 L. L. Ciaccio, (Ed.), Water and Water Pollution Handbook, Vols. 1-4, Marcel Dekker, New York, 1971.
14 D. L. King, in Advances Toward Understanding Lagoon Behavior, Eng. Extension Ser. No. 6, University of Missouri, Columbia, MO, 1966, pp.88-110.
15 D. L. King, Sampling in Natural Waters and Waste Effluents, in L. L. Ciaccio (Ed.), Waste and Water Pollution Handbook, Vol. 2, Marcel Dekker, New York, 1971, p.462.
16 R.D. Cadle, P. L. Magill, A. A. Nichol, H. C. Ehrmantraut and G. W. Newell Sampling Procedures, in C. Ackley, (Ed.), Air pollution Handbook, McGraw-Hill, New York, 1956, Section 10.
17 L. V. Cralley, L. J. Cralley and G. D. Clayton, Industrial Hygiene Highlights, Industrial Hygiene Foundation of America Inc., Pittsburgh, PA, 1968, CH.1.
18 Methods of Air Sampling and Analysis, Amer. Public Health Assoc., Washington, DC 1972, Part 1.
19 R. Perry and R. J. Young (Eds.), Handbook of Air Pollution Analysis, Chapman Hall, London, 1977, Ch.1.
20 R. S. Brief and R. G. Confer, Air Quality Monitoring: Procedures: Data Analysis, in Heating, Piping, Air Conditioning, 1972, pp. 103-110.
21 ASTM Annual Book of Standards, Standard Recommended Practice for Planning the Sampling of the Atmosphere, ASTM, Philadelphia, PA 1973, D-1357-57, Part 23.
22 Air Pollution Manual, Part I, Evaluation, 2nd ed., American Industrial Hygiene Association, Detroit, MI 1972.
23 A. C. Stern, Air Pollution, 2nd ed., Academic Press, New York, 1968.
24 R. J. Charlson, J. Air Pollut. Control Assn., 9 (1969) 802.
25 F. Pooler, J. Air Pollut. Control Assn., 24 (1974) 228-231.
26 H. D.Axelrod and J. P. Lodge, Jr., Sampling and Calibration of Gaseous Pollutants, in A. C. Stern (Ed.), Air Pollution, Academic Press, New York, 1976, Ch.4.
27 Federal Register, 36, (1971), 8195 (Air quality).
28 Federal Register, Part II, Feb. 1978 (Drinking water).
29 G. R. Umbreit, Trace Analysis by Gas Chromatography, in R. L. Grob (Ed.), Modern Practice fo Gas Chromatography, Wiley-Interscience, New York, 1977, p. 417.
30 W. G. Cochran, F. Mosteller and J. W. Tukey, Statistical Problems of the Kinsey Report, American Statistical Assoc., Washington, DC 1954.
31 M. N. Murthy, Sampling Theory and Methods, Statistical Publishing Society, Calcutta, India, 1967, Ch. 1-3.
32 L. Tanner, Probability Sampling of Materials in F. W. Welcher (Ed.) Standard Methods of Chemical Analysis, Vol. 2, Part A, D. Van Nostrand Co., Princeton, NJ, 1963, p.24
33 Acceptance of Evidence Based on the Results of Probability Sampling, ASTM Designation: E141, ASTM, Philadelphia, PA, 1975.
34 Choice of Sampling Size to Estimate the Average Quality of a Lot or Process, ASTM Designation: E122, ASTM, Philadelphia , PA, 1979.
35 W. J. Youden, Statistical Methods for Chemists, Wiley, New York, 1951, Ch. 4,
36 R. B. Dean and W. J. Dixon, Anal. Chem., 23 (1951), 636.
37 W. J. Dixon, Ann. Math. Stat., 22 (1951), 68; Biometrics, 9 (1953) 74.
38 J. F. Fritz and G. W. Schenk, Quantitative Analytical Chemistry, Allyn and Bacon, Boston, MA 1978.

39 G. R. Umbreit, J. Chromatog. Sci., 17 (1979) 482.
40 J. P. Friend, A Report on Research Activities at Drexel University, 1, No. 1 (1978) 5.
41 H. Watson, Amer. Ind. Hyg. Assoc. Quart., 15 (1954) 21.
42 W. C. L. Hemeon and G. F. Haines, J. Air Pollut. Control Assn., 4 (1954) 159.
43 S. Badzioch, J. Inst. Fuel, 33 (1960) 106.
44 V. Vitols, J. Air Pollut. Control Asssn., 16 (1967) 79.
45 E. B. Sansone, Amer. Ind. Hyg. Assn. J., 30 (1969) 487.
46 G. A. Schmel, Amer. Ind. Hyg. Assn. J., 31 (1970) 758.
47 Technical Service Brief TS002, Millipore Corp., Bedford, MA, Feb. 1978.
48 Air Quality Criteria for Sulfur Oxides, Publ. No. AP-50, Cincinnati, OH, (Environmental Protection Agency), 1969.
49 M. Krzymien, Lab. Tech. Rep. LTR-UA (Natl. Res. Counc. Can., Unsteady Aerodyn. Lab), LTR-UA-49, 8pp, 1979; C. A., 91 (1979) 205519b.
50 American Society for Testing and Materials Methods D-1071-55, Part 19, ASTM, Philadelphia, PA, 1971.
51 G. O. Nelson, Controlled Test Atmospheres, Ann Arbor Sci Publ., Ann Arbor, MI, 1971, Ch. 3.
52 R. J. Gordon, H. Maysohn and R. M. Ingels, Environ. Sci. Technol., 2 (1968) 117.
53 D. M. Allen, U. S. Patent, 3,501,899 (1970).
54 G. Herman, Chem. Techn. (Leipzig), 17 (1965) 97.
55 J. P. Lodge, Jr., J. B. Pate, B. E. Ammons and G. A. Swanson, J. Air Pollut. Contr. Assn., 16 (1966) 197.
56 M. Corn and W. Bell, Amer. Ind. Hyg. Assoc., J., 24 (1963) 502.
57 C. Huygen, J. Air Pollut. Contr. Assn., 20 (1970) 675.
58 A. Setterlind, Amer. J. Pub. Health, 34 (1944) 863.
59 H. Scholz and O. J. Stebel, Gas-Wasserfach, 91 (1950) 127; C.A., 44 (1950) 7593d.
60 A. F. Wartburg, H. D. Axelrod, R. J. Teck, M. D. Lattue and J. P. Lodge, Jr., Anal. Chem. 45 (1973) 423.
61 Product Bulletins, Brooks Instruments, Hatfield, PA 1973.
62 Product Bulletins, Tylan Corp., Torrance, CA 1973.
63 W. C. Achinger and R. T. Shigehara, J. Air Pollut. Control Assoc., 18 (1968) 615.
64 W. W. Anderson, Sampling of Solids, in F.W. Welcher (Ed.), Standard Methods of Chemical Analysis, Vol. 2, Part A, Van Nostrand Co., Princeton, NJ, 1963, pp. 28-38.
65 P. M. Giever, Particulate Matter Sampling and Sizing, in A. C. Stern (Ed.), Air Pollution, Vol. 3, 3rd ed., Academic Press, New York, 1976, Ch. 1.
66 J. S. Nader, J. Air Pollut. Contr. Assn., 8 (1958) 35.
67 M. B. Jacobs, The Chemical Analysis of Air Pollutants, Wiley-Interscience, New York, 1960, p. 36.
68 W. J. Smith and A. L. Benson, Spec. Tech. Publ. No. 555, ASTM, Philadelphia, PA, pp. 137-142, 1974.
69 J. B. Littlefield and H. H.Schrenck, U. S. Bur. Rep. Invest. Mines, RI-2392 (1922).
70 A. L. Linch and M. Corn, Am. Ind. Hyg. Assn. J., 26 (1965) 601.
71 H. J. White, Industrial Electrostatic Precipitation, Addison-Wesley, Reading, MA, 1963.
72 L. Waldman and K. H. Schmitt in C. N. Davies (Ed.), Aerosol Science, Academic Press, New York, 1966, pp.137, 162.
73 P. A. Hersch, J. Air Pollution Control Assoc., 19 (1969) 164.
74 D. Jentzch, J. Gas Chromatogr., 5 (1967) 226.
75 Y. Makide, T. Tominaga, T. Rowland and F. Sherwood, Chem. Lett., (1979) 355.
76 F. J. Schuette, Atmos. Environ., 1 (1967) 515.
77 R. A. Baker and D. C. Doerr, Int. J. Air Water Pollut., 2 (1959) 142.
78 C. A. Clemons and A. P. Altshuller, J. Air Pollut. Contr. Assn., 14 (1964) 407.
79 A. P. Altshuller, A. F. Wartburg, I. R. Cohen and S. F. Sleva, Int. J. Air Water Pollut., 6 (1962) 75.
80 T. A. Bellar, J. E. Sigsby, C. A. Clemons and A. P. Altshuller, Anal. Chem., 34 (1962) 763.

81 A. P. Altshuller, T. A. Bellar, C. A. Clemons and E. Vander Zaner, Int. J. Air Water Pollut., 8 (1964) 29.
82 R. S. Ajemian and N. E. Whitman, J. Air Pollut. Contr. Assn., 20 (1970) 310.
83 B. S. Smith and J. O. Pierce, Use of Plastic Bags for Industrial Air Sampling, in Analysis of Industrial Air Pollutants. Vol. III, MSS Information Corp., New York, 1974, p. 94.
84 American Conference of Governmental Industrial Hygienists, Air Sampling Instruments, 4th ed., ACGIH, Cincinnati, OH 1972.
85 American Conference of Governmental Industrial Hygienists, Industrial Ventilation, 13th ed., ACGIH, Lansing, MI 1974.
86 Methods for Determination of Velocity, Volume, Dust and Mist Content of Gases, Bull. WP50, Western Precipitation Corp., Los Angeles, CA 1968.
87 Guide and Data Book 1975 - Fundamentals and Equipment (Annual), Amer. Soc. Heat., Refrig., Air Cond. Eng., New York, 1975
88 D. L. Brenchley, C. D. Turley and R. F. Yarnmac, Industrial Source Sampling, Ann Arbor Press, Ann Arbor, MI 1973.
89 A. Turk, J. I. Morrow and B. E. Kaplan, Anal. Chem., 34 (1962) 561.
90 P. W. West, B. Sen and N. A. Gibson, Anal. Chem., 30 (1958) 30.
91 H. J. Paulus and R. W. Thorn, Stack Sampling in A. C. Stern (Ed.), Air Pollution, Vol. III, Academic Press, New York (1976), Ch.14.
92 G. Pearson, J. W. R. Dutton and F. Elliott, U. K. At. Energy Auth. Prod. Group, PG Rep. 408 (1962); C. A., 58 (1963) 7583a.
93 J. J. Naughton, E. F. Heald and I. L. Barnes, Jr., J. Geophys. Res., 68 (1963) 539.
94 R. Rock and K. Schutz, J. Anal. Chem., 237 (1968) 32.
95 T. Ohira, M. Takeda, T. Ishiguro and O. Koyama, J. Jap.Soc. Air Pollut., 2 (1967) 48.
96 L. Dubois and J. L. Monkman, Mikrochim. Acta, (1970) 313.
97 G. J. Ferguson, Rev. Sci. Instrum. 34 (1963) 403.
98 E. Briner, Helv. Chim. Acta, 21 (1938) 1218.
99 W. Bertsch, E. Anderson and G. Holzer, J. Chromatogr., 112 (1975) 70.
100 B. Versino, H. Knoppel, M. DeGroot, A. Peil, J. Poelman, H. Schauenburg, H. Vissers, and F. Geiss, J. Chromatogr., 122 (1976) 373.
101 M. Dressler, J. Chromatogr. 165 (1979) 167.
102 B. J. Dowty, D. R. Carlisle and J. L. Laseter, Environ. Sci. Technol., 9 (1975) 762.
103 M. J. O'Brien and R. L. Grob, J. Chromatogr. 155 (1978) 129-148.
104 J. Angerer, A. Haag, and G. Lehnert, Comm. Eur. Connunities, EUR 5360, Proc. Int. Sym. Recent Adv. Asses. Health Eff. Environ. Pollut., Vol. 3, 1975, pp 1317-1329.
105 R. Sydor and D. J. Pietrzyk, Anal. Chem., 50 (1978) 1842.
106 A. K. Burnham, G. V. Calder, J. S. Fritz, G. A. Junk, H. J.Svec and R. Willis, Anal. Chem., 44 (1972) 139.
107 J. Kasche and K. Schroeter, Chem. Tech. (Leipzig), 31 (1979) 311; C. A. 91 (1979) 128246f.
108 G. A. Junk, J. J. Richard, M. D. Grieser, D. Witiak, J. L. Witiak, M. D. Arguello, R. Vick, H. J. Svec, J. S. Fritz and G. V. Calder, J. Chromatogr., 99 (1974) 745.
109 W. Wliszczak, F. Meisinger and G. Kainz, Mikrochim Acta, (1971) 139.
110 C. Osterroht, J. Chromatogr., 101 (1974) 289.
111 J. D. Twibell and J. M. Home, Nature, 268(1977) 711.
112 P. R. Musty and G. Nickless, J. Chromatogr., 89 (1974) 185.
113 Gas-Chrom Newsletter, Applied Science Laboratories, Inc. State College, PA Vol. 19, No. 2, March 1978, p.3.
114 V. Leoni, G. Pucetti and A. Grella, J. Chromatogr., 106 (1975) 119.
115 Hewlett-Packard Co., Application Note AN 228-6, Vol. 8, No. 1, 1978.
116 W. Averill and J. E. Purcell, Chromatography Newsletter, Perkin-Elmer Corp., Norwalk, CT 6, No. 2 (1978) 30.
117 J. W. Russell, Environ. Sci. Technol. 9 (1975) 1175.
118 E. Neuber, J. Mueller and R. Sartorius, Mikrochim. Acta, 2 (1979) 131.
119 B. A. Colenutt and S. Thornburn, Chromatographia, 12 (1979) 519.

120 F. A. Bencsath, K. Drysch, D. List and H. Weichardt in M. Eggstein and H. M. Weichardt (Ed.), Mass Spectrom. Comb. Tech. Med. Clin. Chem. Clin. Biochem. Sym., Univ. Tubingen, 1977, pp 421-432; C. A. 91, (1979) 215848x.
121 E. Burghardt and R. Jeltes, Atmos. Environ. 9 (1975) 935.
122 S. K. Yasuda and E. D. Loughran, J. Chromatogr., 137 (1977) 283.
123 J. E. Cuddeback, W. R. Burg and S. R. Birch, Environ. Sci. Technol., 9 (1975) 1168.
124 T. A. Bellar and J. J. Lichtenberg, EPA Report 670/4-74-009, National Environmental Research Center, Cincinnati, Ohio, Nov. 1974.
125 G. Holzer, H. Shanfield, A. Zlatkis, W. Bertsch, P. Juarez, H. Mayfield and H. M. Liebich, J. Chromatogr., 142 (1977) 755.
126 B. V. Loffe, A. Isidorov and I. G. Zenkevich, J. Chromatogr., 142, (1977) 787.
127 M. S. Smith, A. J. Franics and J. M. Duxbury, Environ. Sci. Technol., 11 (1977) 51.
128 E. Schoentube and M. Schaldlich, Ergeb. Exp. Med., 20 (1975) 219; Anal. Abstr., 32 (1977) 3H18.
129 K. Tanaka, K. Fukaya, N. Nishiyama, Y. Wada and S. Fukui, Eisei Kagaku, 23 (1977) 48; Anal. Abstr., 33 (1977) 2H21.
130 J. Charlton, R. Sarteus and J. M. Sharkey, Oil Gas J., July (1975) 96; Anal. Abstr., 31 (1976) 3H24.
131 M. I. Afanas'ev, L. V. Denisova, T. I. Lomovtseva, B. P. Okhatnikov, V. A. Rotin and V. S. Yusfin, Zavod. Lab., 41 (1975) 1202; Anal. Abstr., 30 (1976) 5C44.
132 A. A. Khvastikova, S. A. Reznikov, R. I. Sidorov, L. I. Zuitseva and G. I. Vakhrusheva, Zh. Anal. Khim., 30 (1975) 1001; Anal. Abstr. 30 (1976) 5H16.
133 R. G. Melcher, W. L. Carner, L. W. Severs and J. R. Vaccaro, Anal. Chem., 50 (1978) 251.
134 E. Selke, W. K. Rohwedder and H. J. Dutton, J. Amer. Oil Chem. Soc., 54 (1977) 62.
135 A. Gold, C. E. Dube and R. B. Perni, Anal. Chem. 50 (1978) 1839.
136 M. S. Black, R. P. Herbst and D. R. Hitchock, Anal. Chem. 50 (1978) 848.
137 R. Jeltes, Atmos. Environ. 3 (1969) 587.
138 Y. Hoshika, Taiki Osen Gakkaishi, 14 (1979) 210; C. A., 92 (1980) 10476y.
139 X. Guardino, J. Bartual and M. J. Berenguer, Afindad, 36 (359) (1979) 9, C.A., 91 (1980) 61768g.
140 T. Tanaka, J. Chromatogr., 153 (1978) 7.
141 B. C. Turner and D. E. Glotfelty, Anal. Chem., 49 (1977) 7.
142 J. W. Russell and L. A. Shadoff, J. Chromatgr. 134 (1977) 375.
143 P. Zimmerman and R. Rasmussen, Environ. Sci. Technol., 9 (1975) 1177.
144 M. Yamamura and S. Nishi, Bunseki Kagaku 24 (1975) 635; Anal. Abstr., 30 (1976) 4C76.
145 M. Okumura and S. Nakaniski, Shin. Nippon Denki Giho. 13 (1978) 42; C.A. 91 (1979) 128147z.
146 E. G. Ivanyuk and Y. A. Kolievskaya, Zavod., Lab., 43 (1977) 157; Anal. Abstr., 33 (1977) 2H23.
147 G. O. Wood and R. G. Anderson, Amer. Ind. Hyg. Assoc. J., 36 (1975) 538.
148 F. Guern, LPI Contrib., (1979) 368; C.A. 91 (1979) 186006p.
149 S. M. Rappaport and R. Morales, Anal. Chem., 51 (1979) 19.
150 W. A. Aue and P. M. Teli, J. Chromatogr. 62 (1971) 15.
151 K. Grob, Jr., and G. Grob, J. Chromatogr. 106 (1975) 299.
152 O. Furukawa, T. Hasegawa and Y. Shig ta, Akusku no Kenkyu, 7(34) (1979) 35; C.A. 92, (1980) 10499h.
153 T. Tsugita, T. Imai, Y. Doi, T. Kurata and H. Kato, Agr. Ciol. Chem., 43 (1979) 1351; C.A. 91 (1979) 11222216z.
154 J. Labows, G. Preti, E. Hoelzle, J. Leyden and A. Kligman, J. Chromatogr., 163 (1979) 294.
155 J. S. Parsons and S. Mitzner, Enviorn. Sci. Technol., 9 (1975) 1153.
156 S. G. Zeldes and A. D. Horton, Anal. Chem., 50 (1978) 779.
157 M. Vanhaelen, R. Vanhaelen-Fastre and J. Geeraerts, J. Chromatogr., 144 (1977) 108.
158 M. F. Ives, J. Ass. Offic. Anal. Chem. 58 (1975) 457.
159 R. H. Brown and J. H. Purnell, J. Chromatogr., 178 (1979) 79.

160 I. Kifune, K. Kawata and K. Oikawa, Eisei Kagaku, 25 (1979) 205; C.A. 92 (1980) 115497j.
161 L. Renberg, Anal. Chem., 50 (1978) 1836.
162 J. E. Woodrow and J. N. Seiber, Anal. Chem., 50 (1978) 1229.
163 R. J. Bryant and W. Minett, Pestic. Sci., 9 (1978) 525; C.A., 91 (1979) 152591k.
164 R. S. Braman, J. M. Ammons, and J. L. Bricker, Anal. Chem. 50 (1978) 992.
165 B. J. Tyson, Anal. Lett. 8 (1975) 807.
166 G. Winnett and M. Siewierski, Bull. Environ. Contam. Toxicol., 14 (1975) 681; Anal. Abstr., 30 (1976) 6H23.
167 A. Tateda and J. S. Fritz, J. Chromatogr., 152(2) (1978) 329.
168 Y. Cohen, Anal. Chem., 49 (1977) 1238.
169 Z. Voznakova, M. Popl and M. Berka, J. Chromatogr. Sci., 16 (1978) 123.
170 J. Albert and B. Jonke, Z. Wasser. Abwasser-Forsch., 8 (1975) 140; Anal. Abstr., 31 (1976) 1H63.
171 V. D. Chmil', Zh. Anal. Khim., 30 (1975) 2444; Anal. Abstr. 31 (1976) 3H61.
172 D. P. Beggs, Hewlett-Packard GC/MS Application Note, AN 176-224, 1979.
173 C. A. Burgett and L. E. Green, Hewlett-Packard Application Note, ANGC5-76 (1976).
174 S. N. Chesler, B. H. Gump, H. S. Hertz, W. E. May and S. A. Wise, Anal. Chem., 50 (1978) 805.
175 R. E. Sievers, R. M. Barkley, G. A. Eiceman, R. H. Shapiro, H. F. Walton, K. J. Kalonko and L. R. Field, J. Chromatogr., 142, (1977) 745.
176 L. D. Kissinger and J. S. Fritz, J. Amer. Water Works Assoc., 68 (1976) 435.
177 A. Cavallaro, G. Colli, A. Lemma, G. Invernizzi, L. Luciani and A. Gorni, Boll. Chim. Unione, Ital. Lab. Prov. Parte Sci., 5 (1979) 472; C.A. 91 (1979) 169298t.
178 A. L. Linch, Evaluation of Ambient Air Quality By Personnel Monitoring, CRC, Cleveland, OH 1974.
179 S. Calvert and W. Workman, Ind. Hyg. J., Aug (1961) 318-324.
180 A. Turk, J. Morrow, P. F. Levey and P. Weissman, Int. J. Air Water Pollut., 5 (1961) 14.
181 E. R. Hendrickson in A. C. Stern (Ed.), Air Pollution, Vol. II, Academic Press, New York, 1968, pp. 3-53.
182 M. Feldstein and S. Balestrieri, J. Air Pollut. Contr. Assn. 15 (1965) 177.
183 M. Shepherd, S. M. Rock, R. Howard and J. Stormes, Anal. Chem., 23 (1951) 1431.
184 E. R. Weaver, E. E. Hughes, S. M. Gunther, S. Schuhmann, N. T. Redfearn and R. Gordon, J. Res. Nat. Bur. Stand., 59 (1957) 383.
185 G.C.B. Carie, Tappi 46 (1963) 1.
186 H. Schmidt and C. E. Doering, Chem. Tech. (Leipzig) 288 (1976) 677; Anal. Abstr., 32 (1977) 6J30.
187 M. Deimel and W. Dulson, Staub-Reinhalt. Luft, 39 (1979) 332; C.A., 92 (1980) 46671x.
188 J. Angerer, A. Haap and G. Lehneri, Z. Klin. Chem. Klin. Ciochem., 13, (1975) 129; Anal. Abstr., 29 (1975) 5H29.
189 D. R. Cronn and D. E. Harsch, Anal. Lett. 9 (1976) 1015.
190 Y. Hoshike, Taiki Osen Gokkaishi, 14 (1979) 210; C.A. 91 (1980) 10476y.
191 W. V. Cropper, Sampling of Liquids, in F. J. Welcher (Ed.), Standard Methods of Chemical Analysis, Vol. 2. Part A. Van Nostrand Co., New York, 1963, p.46.

RELATED READINGS ON SAMPLING

1 An Investigation on Characterization of Trace Odorants in Air by Gas Chromatography Using Preconcentration; Its Evaluation and Application. Y. Hoshika and G. Muto, Bunseki Kagaku, 29(2)(1980)T10-19.

2 On-Column Cryogenic Trapping of Sorbed Organics for Determination by Capillary Gas Chromatography. D. Kalman, R. Dills, C. Perera and F. DeWalle, Anal. Chem., 52(12)(1980)1993-4.

3 Solid Sorbent Collection and Gas Chromatographic Determination of the Propylene Glycol Butyl Ether Esters of 2,4,5-T in Air. M.L. Langhorst, Am.Ind. Hyg. Assoc. J., 41(5)(1980)328-33.

4 Base Peak Profiles of Gas Chromatography-Mass Spectrometric Data Obtained From Thermal Desorption of Activated Carbons. K. Alben, Anal. Chem., 52(12)(1980)1821-4.

5 Field Comparison of Polyurethane Foam and Tenax-GC Resin for High-Volume Air Sampling of Chlorinated Hydrocarbons. W.N. Billungs, T.F. Bidleman, Environ. Sci. Technol., 14(6)(1980)679-83.

6 Sampling of Hydrocarbons in the Atmosphere for Gas Chromatographic Analysis. V. Simik, M. Jednacak, Nafta(Zagreb), 31(5)(1980)262-6.

7 Determination of the Pollution of the Bay of Tallinn by Petroleum Hydrocarbons using Infrared Spectroscopy and Gas Chromatography. K. Tika, A. Talvari, Kh.I. Yankovskii, Tezisy Dokl.-Resp. Konf. Molodykh Uch.-Khim., 2nd, 2(1977)28.

8 Automation of Gas Chromatographic Analysis of Air Samples Trapped on Activated Carbon Ampuls. M. Ferm, U. Samuelsson, Inst. Vatten-Luftvardsforsk., (Publ.) B, IVL B559(1980)35 pp.

9 Methods of Evaluation of Potentially Hazardous Organics in Water. R.P. Collins, Report, W80-02761, OWRT-A-072-CONN(1)(1980). Order No. PB80-135122, Gov. Rep. Announce, Index(U.S.), 80(11)(1980)1844.

10 Method for Sampling a Liquid for Gas Chromatography. Hokushin Electric Works, Ltd., Jpn. Kokai Tokkyo Koho 80, 101,046,Aug. 1, 1980.

11 Houston Aerosol Characterization-GC/MS Study. C.M. Sparacino, Spec. Conf. Ozone. Oxid. Interact. Total Environ., (Proc.), 2nd(1979)234-46.

12 Analytical Methods and Their Problems in the Analysis of Oil in Water. R.J. Law, Oily Water Discharges: (Proc. Semin.)(1978)(Publ. 1980), 167-76. Edited by C.S. Johnston and R.J. Morris, Appl. Sci., Barking, Engl.

13 Behavior of Hydrocarbons in the Atmosphere. 1. Continuous Sampling with Bags for Determination of Aliphatic Hydrocarbons. A. Matano, A. Kemmotsu, T. Mizuhata and K. Ishii, Okayama-ken Kankyo Senta Nempo, 4(1980)62-8.

14 Behavior of Hydrocarbons in the Atmosphere. 2. Sorption with Tenax Beads for Determination of Aromatic Hydrocarbons. A. Matano, A. Kemmotsu, T. Mizuhata, and K. Ishii, Okayama-ken Kankyo Hoken Senta Nempo, 4(1980)69-74.

15 Reactive Adsorbent Derivative Collection and Gas Chromatographic Determination of Chloromethyl Ether in Air. M.L. Langhorst, R.G. Melcher, and G.J. Kallos, Am. Ind. Hyg. Assoc. J., 42(1)(1981)47-55.

16 Sampling and Intake of Total Atmospheric Hydrocarbons for Gas Chromatographic Determinations. Rong-Xian, Zhang,Huan Ching K'o Hsueh, 2(1)(1981)61,18.

17 A Gas-Chromatographic Method for the Determination of Vinylidene Chloride in Air. K.S. Kirpal, J. Anal. Toxicol. 4(5)(1980)266-8.

18 Gas Chromatographic Determination of Vinyl Chloride and Vinyl Acetate in the Air at Work Places. M. Dobecki and J. Krajewski, Chem. Anal. (Warsaw), 25(3)(1980)351-7.

19 Evaluation of the Basic GC/MS Computer Analysis Technique for Pollutant Analysis. J.E. Bunch, N.P. Castillo, D. Smith, J.T. Bursey and E.D. Pellizzari, Gov. Rep. Announce. Index(U.S.), 80(22)(1980)4744.

20 Determination of Nitroglycol in the Ambient Air by Gas Chromatography. S. Tanaka, A. Akabane and Y. Urashima, Sangyo Igaku, 22(5)(1980)390-1.

21 Method for Trapping Low Levels of Phosphine at Ambient Temerature for Gas Chromatographic Analysis. T. Dumas, E.J. Bond, J. Chromatogr., 206(2)(1981)384-6.

22 Collection and Analysis of Trace Organic Vapor Pollutants in Ambient Atmospheres. The Performance of a Tenax-GC Adsorbent Tube. R.H. Brown, C.J. Purnell, J. Chromatogr., 178(1)(1979)79-90.

23 An Application of GC/MS to the Analysis of Odorants in Air. S. Nakayama, T. Ishiguro, Y. Shigeta and T. Kato, Koenshu-Iyo Masu Kenkyakai, 4(1979) 121-30.

24 Gas Chromatographic Analysis of Acetic Acid in Air. K.E. Williams, J.F. Mazur, Am. Ind. Hyg. Assoc. J., 41(1)(1980)1-4.

25 Gas Chromatography-Mass Spectrometric Identification of Organic Compounds and Determination of Acrylonitrile in the Air of the Kawasaki Industrial Area. H. Suzuki, S. Satoh, Koensho-Iyo Masu Kenkyukai, 4(1979)113-20.

26 Studies on Offensive Odor. 2. Analysis of Odorous Substances from Solid Wastes by Gas Chromatography-Mass Spectrometry. A. Kamiya, H. Mizutani, M. Hayakawa, D. Aoyama and Y. Osi, Kankyo Osen Busshitsu to Sono Tokishikorojii Shinpojumu, (Koen Yoshishu), 6th(1979)43-5.

27 Sorption of Phenols from Water and Subsequent Thermal Desorption for GC Analysis. Z. Voznakova, M. Popl, J. Chromatogr., 17(12)(1979)682-6.

28 Method for the Determination of Toxic Substances in the Air Using Chromatography and Mass Spectrometry. M.T. Dmitriev, E.G. Rastyannikov, S.A. Volkov, A.G. Malysheva and B.M. Samokhodkin, Gig. Sanit., (5)(1980)42-4.

29 Detection of Environmental Pollutants. Measurements of Nitrosoamine in Atmosphere. K. Aihara, and M. Sinozaki, Kanagawa-ken Kogai Senta Nempo, 10(1977)184.

30 Sampling for Chemical Analysis. B. Kratochvil and J.K. Taylor, Anal. Chem., 53(1981)925A.

CHAPTER 4

SAMPLE TREATMENT

4.1 INTRODUCTION

Sample treatment is that process of manipulative procedures used by the analytical chemist to prepare the sample for the analytical measurement stage. Sampling per se is not performed in the laboratory; sampling is the process external to the laboratory environment (see Chapter 3) which is necessary for the analytical chemist to judge whether or not the sample is truly representative.

The technique of headspace gas sampling is discussed in Section 4.3.2.1. This is a subtopic of the general section dealing with extractions. Although it is referred to as a sampling technique, basically it is a sample treatment procedure. The justification for this statement can be seen in the fact that if the sample is not representative at this point, great attention to procedural detail will not produce data which the analytical chemist can consider as truly representative.

4.2 STORAGE OF SAMPLES

Chemical preservatives should be added to samples for chemical, physical, bacteriological or radiological examination only when specified in the particular test method. Quick freezing may be beneficial in preserving some organic constituents.

If the samples are to be held more than 1 h before examination, they should be stored in a refrigerator or ice chest at a temperature of not more than 4°C. The interval between collection and analysis of water samples should not exceed 12 h, if at all possible. Shipment of the sample should be in an insulated iced container to maintain the temperature between 0° and 4°C to permit analysis within the 12 h period.

This is one of the steps in the series from the sampling stage to actual determination of a component in an environmental sample which often is overlooked. Samples are collected, transported to the laboratory, only to wait for various time intervals (hours to days) before the actual analysis. Environmental samples emphasize a very important problem which many chemists may forget: one is not always sure of all the components in the sample. There may be components present which are photochemically active which may interact with other sample components or which simply undergo some rate-controlled chemical decomposition with time.

A good rule of thumb is to assume that the sample received (gas or liquid) will undergo some type of change with time and, thus, take proper precautions to minimize these changes. Storage in the dark at a low temperature (at least 4°C) is possibly the best procedure one could practice.

Mold can grow in water samples, and therefore the low temperature storage will reduce this difficulty. If analysis will not be performed within a reasonable time interval (1-2 days), adequate storage space must be available in the laboratory. Sample containers which have not been chosen correctly and thoroughly cleaned will adsorb a number of the contaminants on the walls.

It sometimes is necessary in air pollution studies to determine whether a pollutant has been adsorbed or absorbed by the local vegetation. If this be the case, fresh plant samples must be protected from enzyme action, e.g., autolysis and/or fermentation. Mold growth on such samples can cause errors in subsequent analyses. Samples which will be analyzed within a day or two may be kept in glass jars which are tightly sealed. If more than a couple of days will pass before analysis is possible, these types of samples should be stored in the frozen state (ref. 1). Once thawed, they must be analyzed as decomposition rate will be accelerated. Miles et al. (ref. 2) described a technique for the storage and analysis of water and mud samples contaminated with Abate. Giever (ref. 3) has noted that environmental samples should be stored away from heat and light and the container material checked for interactions with the sample. He recommends glass containers over synthetic compound containers.

Samples which must be diluted prior to analysis should not be diluted until time for the analysis. Decrease of contaminant concentration because of adsorption effects will be more pronounced from a dilute solution than from a concentrated solution. It has been shown (ref. 4) that storage of water samples containing Cl_2 or Br_2 increases the concentration of haloalkanes (e.g., chloroform, bromodichloromethane, dibromochloromethane and bromoform) because of reactions between halogen and haloalkanes. Ascorbic acid, sodium sulfate, sodium bisulfite, sodium thiosulfate or hydrazine will stabilize the solution.

4.3 EXTRACTION TECHNIQUES

Extraction of environmental samples may be classified by the sample type (i.e., physical state of sample).

(1) Leaching (liquid-solid equilibrium). Extraction of components of interest from a solid sample by use of a liquid-extracting solvent. The analysis of particulate samples for polynuclear aromatic hydrocarbons (PAHs) is an example of this type of extraction system.

(2) Liquid-liquid partition (liquid-liquid equilibrium). The extraction of pollutants from a liquid (organic or aqueous) sample by use of another liquid phase which is immiscible with the sample phase. An example of this technique is the extraction of organic pollutants dissolved in a water sample, i.e., octane extraction of waste water effluents or water extraction of ionic species (ion pairs) from petroleum-base samples.

(3) Gas-liquid partition (gas-liquid equilibrium). This type of system can have several subdivisions:

(a) Headspace sampling for volatile contaminants from an aqueous sample. An example would be the headspace analysis of low molecular weight alkyl halides from water samples (e.g., ethyl chloride or vinyl chloride).

(b) Water extraction of a gas sample for water-soluble components, e.g., extraction of SO_3 or CO_2 from air samples or aerosol extraction with water.

(c) Organic solvent extraction of a gas sample, e.g., lyophilic aerosols in an air sample.

(4) Gas-solid equilibration (gas-solid equilibrium). Sampling of volatile materials from a solid with an inert (He or N_2) gas, e.g., vapor equilibration (headspace) of vinyl chloride volatiles from a polymeric substance or odors from a painted surface.

(5) Extraction of liquid by a solid (solid-liquid equilibrium). This is referred to by a number of workers in the area of environmental analysis as an extraction technique (which it rightfully is); however, following our delineation between sampling (external to the laboratory) and sample treatment (internal to the laboratory), this technique is discussed in Chapter 3, Sections 3.4.1 and 3.4.2. The use of carbon as the adsorbent (CAM) is a standard technique from which the carbon-chloroform extract (CCE) and carbon-alcohol extract (CAE) are obtained (refs. 5-9). Dressler (ref. 10) has provided an excellent overview of this technique.

(6) Extraction of air by a solid (solid-gas equilibrium). See the discussion in classification number 5 above.

The first four classifications have one basic process in common, i.e., equilibration of two phases. Thus, they all involve an equilibrium constant K_D which is dependent upon temperature and pressure. Each system may be defined by the Nernst distribution law (ref. 11).

$$\underset{\text{Phase 1}}{\text{Component S}} \rightleftharpoons \underset{\text{Phase 2}}{\text{Component S}} \tag{4.1}$$

$$K_D = [S]_2/[S]_1 \tag{4.2}$$

Eqn. 4.2 is valid only when S is the same species (i.e., has the same molecular weight) in both phases. If component S has a different form in one of the phases, a better description of the equilibrium would be the distribution ratio D.

$$D = \frac{[\text{total concentration of S (all forms) in phase 2}]}{[\text{total concentration of S (all forms) in phase 1}]} \tag{4.3}$$

The total concentration of the forms of S is the analytical concentration C_S. The analytical concentration of a substance is equal to the sum of the equilibrium concentrations, thus

$$(C_s)_2 = \sum_{i=1}^{n} [S_i]_2 \qquad (4.4)$$

$$(C_s)_1 = \sum_{i=1}^{n} [S_i]_1 \qquad (4.5)$$

We may then rewrite Eqn. 4.3 as

$$D = (C_s)_2/(C_s)_1 \qquad (4.6)$$

The use of eqn. 4.6 only becomes important when one has ionic species, metal ion being removed by chelation or when dimerization or polymerization reactions are taking place. Most importantly it is used when extracting high concentrations of phenols or metals, e.g., in the Consent Decree List of Priority Pollutants. However, most environmental samples have low concentrations of phenols and metals; therefore the K_D value is of more importance than the D value. Additionally, when dealing with phenols for extractions we work with very acidic solutions so the undissociated form predominates, i.e., the equilibrium is driven to the left:

$$C_6H_5OH \underset{}{\overset{H^+}{\rightleftharpoons}} C_6H_5O^- + H^+ \qquad (4.7)$$

The use of metal chelates for extractions of metals is simplified by using an excess of the chelating reagent and driving the equilibrium to the right:

$$M^{n+} + nL(xs) \underset{K_f}{\rightleftharpoons} ML_n \qquad (4.8)$$

One should realize that what is presented in eqns. 4.7 and 4.8 represents dynamic equilibria and that a small amount of the component of interest will remain in the aqueous phase. However, the quantity remaining is a very small amount of the original concentration and in many cases will introduce little or no error in the analytical determinations. Some components will not have a distribution constant high enough to permif quantitative extraction in one transfer. Thus, we find ourselves faced with two options: (1) Carry out multiple extractions with fresh aliquots of solvent and measure the amount of pollutant in each aliquot, or (2) Perform one extraction and make a correction for amount unextracted.

The second option can be risky if one is not sure of the complete sample matrix. If the distribution constant is known and no other components (present in the sample) interfere, the amount remaining unextracted can be calculated. Another consideration is whether it exists in more than one form. Some organic compounds may dimerize or polymerize in the organic layer (benzoic acid dimerizes in benzene) or they may form an ion association product (amines could be extracted into the organic layer by ion-pairing with an anion from an acidic aqueous solution).

4.3.1 Liquid-liquid extraction

For clarity we will divide this topic into two subheadings: (1) macro-extractions where large volumes of sample (up to several liters) and extracting solvent (50-200 ml) are uesd; (2) micro-extractions where small volumes of sample (10-25 ml) and extracting solvent (0.2-3 ml) are used. Unfortunately most of the EPA methods of analysis employing liquid-liquid extraction use the macro-extraction method (ref. 12). This is presently being revised (ref. 13).

4.3.1.1 Macro-liquid-liquid extractions. Macro-extractions are not attractive because they require large sample volumes (not always available), require a large volume of the extracting solvent, introduce errors (sample component losses) during the concentrating procedure (usually an evaporative technique), and are costly. These macro-extraction procedures evolved primarily because the concentration levels of the pollutants were low and the methods of detection lacked sensitivity. This situation has been reversed to a large extent in the case of many of the priority pollutants. This has been brought about by the increased sensitivity and selectivity of many of the present-day detector systems (see Chapter 5, Section 5.5).

Purity of extracting solvent is of concern when used for separation of environmental sample components. Solvent contaminants may be of the same order of magnitude as the sample component of interest. Not only can this cause spurious peaks but it may enhance the peak one is monitoring. Choice of the wrong solvent may result in a large solvent peak which overlaps the peak(s) of interest. The ideal extraction solvent would be one that is very selective for types of compounds, has a very low boiling point and has a retention volume equal to the void volume of the column, i.e., $V_M = V_R$.

Methylene chloride (CH_2Cl_2) possesses many of the properties of an ideal extracting solvent, providing one is not analyzing for halogenated contaminants which have low boiling points. It boils at 35°C and has a density of 1.336 g/ml. This latter property makes it an excellent solvent because it will be the bottom layer in an aqueous-methylene chloride system, thus providing ease of removal from the extracting vesesl. Some of the other commonly used extraction solvents (n-C_5-, n-C_6-, and n-C_8-hydrocarbons) have densities less than water and will constitute the top layer in the extraction system.

Benzene has been recommended in some of the older literature; but, because of its toxic effects, it should be avoided. n-Pentane and n-hexane usually can be substituted with little or no decrease in extraction efficiency.

The question of how many extractions are necessary to be certain of quantitative transfer to the organic phase is a valid one. If we knew the distribution constant of every pollutant, in any type of matrix, this would be easy to answer. We can, however, calculate the fraction remaining f_n by use of eqn. 4.9.

$$f_n = \left[\frac{1}{1 + K_D\left[\frac{V_o}{V_w}\right]}\right]^n = \left[1 + K_D\left[\frac{V_o}{V_w}\right]\right]^{-n} \quad (4.9)$$

where n = number of extractions or transfers,

K_D = distribution constant,

V_o = volume organic phase,

V_w = volume aqueous phase.

Unfortunately, we do not know the distribution constants of many of the priority pollutants, but we are making progress.

To demonstrate the effect of a correct extracting solvent, let us present a few calculations using eqn. 4.9. Hites (ref. 14) used as a typical example of a macro-extraction one in which V_w = 3.5 ℓ and V_o = 200 ml. Taking his volumes, let us calculate:

(1) fractions remaining after 1-10 extractions where the K_D = 1 (contaminants equally soluble in both phases);

(2) fractions remaining after 1-10 extractions where the K_D = 5 (contaminants 5X more soluble in organic phase than aqueous phase);

(3) the value of K_D necessary to extract 99%, 95% and 90% of the contaminant using one equilibration (extraction). This would compare to usual extraction conditions of large water samples.

The results of these calculations are shown in Table 4.1.

May and Stalling (ref. 15) have described the use of double-reservoir rotoevaporation vessels (250 ml and 100 ml) for solvent evaporation after methylene chloride extraction, gel permeation and liquid chromatography clean-up steps. This technique circumvents the problems associated with solvent transfer and organic solvent concentration (e.g., Kuderna-Danish apparatus) steps. Petroleum products in natural water have been identified and determined by gas chromatographic analysis of the hexane extract using a SE-30 or OV-1/Gas-Chrom Q column and flame ionization detection (ref. 16). BHC residues were determined by GC analysis of the extracts from freshwater in Northern Ireland (ref. 17). They were able to detect 1 mg/ℓ of benzene hexachloride and 1-5μg/ℓ (ppb) of Dieldrin, p,p'-DDT and its metabolites. Overton et al. (ref. 18) separated and determined petroleum hydrocarbons in marine sediments using wall-coated open tubular glass capillary columns. Phenols in water were extracted with diethyl ether (ref. 19), converted to their fluorinated derivatives and detection was by electron capture. Thiophene (and chlorinated derivatives), benzene and toluene have also been extracted from waste water and determined in the extracted distillates from soil, water and plants (ref. 21). A thermionic detector was used and

TABLE 4.1

Example data of extraction efficiency for water pollutants

V_o = 200 ml; V_w = 3.5 ℓ.

(1) Fraction remaining unextracted: K_D=1	(2) Fraction remaining unextracted: K_D=5
f_1 = 0.946	f_1 = 0.778
f_2 = 0.895	f_2 = 0.605
f_3 = 0.847	f_3 = 0.471
f_4 = 0.801	f_4 = 0.366
f_5 = 0.758	f_5 = 0.285
f_6 = 0.717	f_6 = 0.222
f_7 = 0.678	f_7 = 0.173
f_8 = 0.641	f_8 = 0.134
f_9 = 0.607	f_9 = 0.104
f_{10} = 0.574	f_{10} = 0.081

(3) n = 1

% Extracted	K_D-Necessary
99	1734
95	333
90	158

a detection threshold of 0.02 mg/kg was obtained. Organophosphorus pesticides have been extracted from water samples (refs. 22 and 23) and determined by a method yielding a detection limit of 1 ppb and an efficiency of > 95%. The study (ref..22) showed that pesticides were adsorbed on the walls of the syringe from hexane extracts but not from acetone extracts. Stalling and Huckins (ref. 24) analyzed and characterized toxaphene in fish and water by an extraction technique followed both by electron-capture detection and gas chromatography-mass spectrometry. Carbon-14 labeled pesticides were extracted from serum with hexane, separated and determined using capillary columns coated with SE-30 or OV-1 (ref. 25). Flunitrazepam, desmethylflunitrazepam and clorazepam were extracted from plasma and the final solution was a hexane-acetone (4:1) mixture. This was separated and analyzed on an OV-225 column using an electron-capture detector (ref. 26). Aldicarb, aldicarb sulfoxide and aldicarb sulfone have been extracted from water with chloroform, separated on a Florisil column and detected with a Hall electrolytic detector (ref. 27). Sensitivity of the method is 0.05 ppm aldicarb. Smoke condensate flavorings have been extracted from ground sausage with diethyl ether and determined by gas chromatography (ref. 28). One-tenth ppm could be detected. Ethylenethiourea was extracted and derivatized in one step from water samples (ref. 29). Electron-capture and flame-ionization detection were used for identification and quantification.

4.3.1.2 Micro-liquid-liquid extractions. The advantages of the micro-extraction technique are that one is able to extract and concentrate pollutants simultaneously and there is no problem with subsequent evaporations. In most macro-extraction methods, the CCE (carbon-chloroform extract) and the CAE (carbon-alcohol extract) techniques, the solvent phase requires further concentration which usually results in serious losses of sample components.

Murray (ref. 30) developed an extraction flask similar to the rapid liquid extraction technique used by Grob et al. (ref. 31). Fig. 4.1 is a representation of this flask wherein V_o = 200 μℓ and V_w = 980 mℓ. Murray compared this micro-extraction method to two macro-methods (steam distillation extraction and a continuous extraction system). To show the loss of solute during the evaporation step, standards of n-hexane were made to contain 1.0 ppm of a group of organic compounds: pesticides, n-alkanes and phthalates. A 10.0 ml aliquot of this standard was then evaporated to 1 ml and each component in the standard was determined by gas chromatography. Two evaporative techniques (micro-Snyder column and rotary evaporator) were used. The recoveries ranged from 37-84% with the micro-Snyder technique and 8-100% with the rotary evaporator.

A comparison of the resulting chromatograms from the micro-extraction and the two macro-methods showed that the micro-technique yielded a better recovery of the spiked solution. The data showed a concentration factor of 1250X for the micro-extraction technique compared to the macro-technique.

To emphasize the advantage of a micro-extraction, let us present the calculations for a theoretical system, the results of which are in Table 4.2. As the data in the table point out, less has been extracted in the 0.200-ml (200 μl) solvent volume; however, the detector would sense more mass of the pollutant than in the other cases. Thus, the advantages one finds are that: (1) the signal from the detector is the same order of magnitude as if one had extracted with a larger solvent volume (i.e., the chromatograms look the same) and (2) the evaporation step and its allied loss of sample were by-passed.

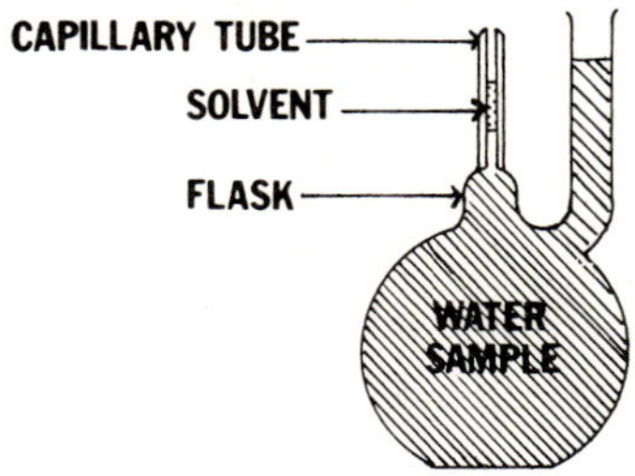

Fig. 4.1 Representative micro-extraction flask.

TABLE 4.2

Comparison of macro- and micro-extraction techniques

Sample: 1ℓ of waste water containing 1 mg/ℓ of contaminant C, i.e., 1 ppm. Solvent n-hexane. Final volume of solvent after evaporation: 1.00 ml (except 200 μl extract). K_D of contaminant C = 15.

Volume extracting solvent: n-C_6(ml)	C remaining(mg)	C extracted* (mg)	Loss from evaporation (%)	C in final volume(mg)	Sample volume injected (μℓ)	Quantity of C injected** (mg)	Amount of C injected*** (mg)
100.0	0.4000	0.6000	50	0.3000	50	0.015	0.0003
10.0	0.8696	0.1304	50	0.0652	50	0.00326	0.000652
5.0	0.9302	0.0698	50	0.0349	50	0.001745	0.000698
0.200	0.9970	0.0030	0	0.0030	50	0.00075	0.00075

*$W_{remaining} = W_{initial}\,(V_w/(K_D V_o + V_w))$.

** Extra digits carried to demonstrate point.

*** Amount injected in a 50 μl sample from original extract, i.e., no evaporation.

The micro-extraction technique has been applied to several of the EPA priority pollutant methods with a high degree of success. A recent paper by Rhoades and Nulton (ref. 32) has demonstrated this technique for the extraction of aromatics, phthalates, polynuclear aromatics and phenols.

4.3.2 Gas-liquid extraction

In this system for sample treatment the solvent for the water pollutants is a gas (clean air, nitrogen or helium). The pollutants which may be extracted by this technique must have a high enough partial pressure (P_i) that they can easily distribute between the gas and liquid phases, i.e., their partial pressure (in terms of moles) should exceed their water solubilities (expressed in moles/liter). There are few data available on the solubilities of the hydrocarbon types of priority pollutants. Several reasons can account for this discrepancy in the literature: (1) Hydrocarbons and hydrocarbon types are relatively insoluble in water and prior to environmental regulations they were usually classified simply as "insoluble". Due to these regulations, in many countries these concentrations had to be calculated because most pollutants are present at the ppm or ppb levels. (2) No one really needed the data so they were never determined. (3) In the past, analytical techniques were not available to determine these low concentrations.

Several techniques are available under the general heading of gas-liquid extractions.

(1) Headspace analysis or vapor equilibration (Section 4.3.2.1). This is a true equilibrium system. A volume of inert gas is equilibrated with a known volume of water sample (refs. 33-36).

(2) Gas stripping or purge-and-trap technique (Section 4.3.2.2). This is a nonequilibrium technique. The water sample is sparged with an inert gas and the gas phase collected by adsorption on a solid (see Section 3.4), absorbed by a trap containing a solvent in which gases of interest are absorbed (see Section 3.7) or condensed in a trap which is cooled below the b.p. of the lowest boiling gaseous component. In this latter case the cold trap may contain a solid adsorbent or simply glass wool (ref. 37-42).

(3) Gas-liquid extraction from a water sample to which has been added a sufficient quantity of a soluble salt (e.g., KCl, Na_2SO_4, NaCl, etc.) to increase the ionic strength of the sample (this is really a modification of technique No. 1 (see Section 4.3.2.3)). The presence of the salt decreases the solubility of nonpolar and/or slightly polar organic solutes, thus allowing more vapor to enter the gas phase (refs. 43 and 44).

Each of these will be discussed in subsequent sections. Gas-liquid extraction is being used more and more in the area of environmental analysis. It will soon have more direct application to waste water effluents. Many organic molecules which were not previously determined by this technique are now easily measured, because of the advantages (see Section 3.9) and the sensitivity of detectors (e.g., flame ionization and electron-capture detectors).

Kojima (ref. 45) studied the pungent odors from the hydrolyzate of Wasabia japonica. Components identified and determined were allyl isothiocyanate, isopropyl-iso-thiocyanate and 3-butenyl isothiocyanate.

Reshetnikova et al. (ref. 46) sampled the volatiles (b.p. 20-300°C) from coatings during setting. Compounds such as CO, acetaldehyde, acetone, toluene, benzene, butyl alcohol, butyl acetate, isobutanol and ethyl benzene were identified. Sensitivity of method is given as 0.05-0.1 mg/m^3. Nikitin et al. (ref. 47) determined hydrocarbons in air, snow, tree bark and leaves by placing the sample in a saturated sodium chloride solution and heating to 80-90°C. The volatile hydrocarbons were swept out of the containers with carrier gas. Fedchenko and Vigdergauz (ref. 48) showed the effect of water vapor as eluent as compared to nitrogen. Compounds studied were the C_5-, C_8-, and C_{10}-acid nitriles; C_4-, C_7-, and C_9-alcohols and C_8-, C_{11}- and C_{13}-alkanes. The HETP (height equivalent to a theoretical plate) values decreased with the transition from nitrogen to water vapor. Elution times were 2.4X longer with nitrogen.

Methane, ethylene and nitrous oxide were identified in the headspace vapor above soil samples using this technique. The volatiles were separated and determined using both a Porapak N and a Porapak S column (ref. 49). Volatile fatty acids (plus iosmers up to C_5) in urine samples were determined on a 3% Fluorad FC 430/Chromosorb G column. The carrier gas was nitrogen and the detection was by flame ionization (ref. 50). Headspace sampling of the volatiles over roasted coffee (refs. 51 and 52) was used to study the aging process. Nitrogen carrier gas and flame ionization and flame photometric detector (the latter is specific for sulfur compounds) were used. Volatile materials decreased more in quantity with time for the whole-roasted bean than for the ground-roasted beans. Using the headspace technique, Rath et al. (ref. 53) were able to determine CO_2 in HCl gas at the 5-100 ppm level. Hydrogen gas was the carrier and TCD was the means of detection. Volatiles in beer were determined by the headspace technique (ref. 54) by means of adding 28 g of NaCl to 100 ml of beer and heating in a closed container to 40°C. Diketones, high-molecular-weight alcohols and esters were determined using electron-capture and flame ionization measurements.

Tarasova and Kataeva (ref. 55) determined vinyl chloride in polymeric substances. The polymeric sample was placed in water, aqueous alcohol or sunflower oil added and the container sealed. The contents were heated, sampled by syringe and separated on a Apiezon L/Chromosorb W column. Drozd and Novak (ref. 56) have written a review of headspace gas analysis. Various aspects of the technique are discussed as well as a survey of applications. The technique has been used to determine chlorinated hydrocarbons in water samples (refs. 57 and 58). Volatile halothane metabolites from biological tissue have been studied by Maiorino et al. (ref. 59). They were able to detect 2 pmoles/ml in blood and 10 pmoles/g in liver samples. Aromatic hydrocarbons in gases saturated with water vapor have also been done by this technique (ref. 60). These same authors determined the K_D values for benzene, toluene and m-xylene between water and air at temperatures ranging from 10-30°C. Kolb et al. (ref. 52) utilized the technique for samples injected onto capillary columns. This eliminated the need for inlet splitters and inlet pressure was regulated with narrow bore capillaries. Their system was used for analysis of flavors, water pollutants and crude oil vapor extracts. Vinyl chloride and dichloroethane in ethylene glycol were determined in the headspace from sealed containers at 30°C (ref. 61). The sensitivity of the method is 5 x 10^{-5} volume percent with a relative error $\leq$ 5%. Benzene, toluene and xylene were measured in air samples collected in acetic acid (ref. 62). The acid solution was made basic with potassium hydroxide and the headspace vapors analyzed by gas chromatography. The standard deviation of the technique is 0.02-0.04 mg/m^3. Nitrous oxide has been determined by headspace sampling and electron capture detection (ref. 63). Precision better than 2% was achieved with water samples $\geq$ 60 ml. Long-term storage of samples showed a 2.3% variation in concentration of the nitrous oxide. A study of headspace sampling by standard addition for the determination of benzene in water at the 1-1000μg range in a container with 50 ml of air and 50 ml of water, showed 8-2% error at the respective levels (ref. 64). These levels correspond to 0.02-2 ppm benzene in the aqueous phase. A reverse use of headspace sampling was the determination of water in sincalide at 100°C (ref. 65). The water vapor was determined by gas chromatography using a Porapak Q column and thermal conductivity detection. Hydrogen sulfide in heated milk was determined by gas-liquid chromatographic headspace analysis (ref. 66), also. The headspace sample was separated on a polytetrafluoroethylene column of poly(phenyl ether) and phosphoric acid and detected by flame ionization.

The headspace technique has been applied to analysis of vinyl chloride in vinegars and alcoholic beverages (ref. 67), wine aromas (ref. 68, acrylonitrile monomer in plastic packaging and beverages (ref. 69) and trace organics (benzene, monochlorobenzene and dichlorobenzene) in HCl gas, methylene chloride and methylene bromochloride in HBr gas (ref. 70) and volatile aromas from the gardenia flower (ref. 71). Smith et al. (ref. 72) converted hexafluoroacetone from industrial waste water to fluoroform (CHF_3) for headspace analysis by gas chromatography. The method was linear for CHF_3 concentration of 10 ppb-1000 ppm.

4.3.2.1 <u>Headspace and/or vapor equilibration</u>. Headspace analysis is a method of determining volatile organic components as pollutants in water in which the aqueous phase is equilibrated with an inert gas phase usually of equal volume. The fundamentals of this technique have been described by McAuliffe (ref. 35).

The method may be defined as the extraction of components of interest by a gas and the subsequent analysis of this gas phase by gas chromatography. This technique may broadly be separated into two classifications:

(1) The analysis of the vapor phase, in equilibrium with the liquid phase, from a closed static system;

(2) The sparging of the liquid phase by an inert gas followed by analysis of the gas phase. In this classification, the gas phase may be trapped by adsorption or condensed at cryogenic temperatures (refs. 73-77).

In the first classification, the concentrations of the components in the gas phase do not change appreciably unless a large sample has been withdrawn. In the second classification, the component concentrations decrease continuously with time. Thus, the system deviates from equilibrium; in fact the component concentration approaches zero asymptotically.

Difficulties may arise if this technique is used improperly, both qualitatively and quantitatively. One could not expect a complete qualitative analysis of a commercial sample by this technique because many of the sample components may only be present at the ppm level or even ppb level. Bassette et al. (ref. 78) selectively removed components from the sample by use of specific chemical reagents. For example, they were able to remove ketones from the equilibrated gas phase by reaction with acidified hydroxylamine reagent and esters with alkaline hydroxylamine reagent. Sulfur-containing compounds were selectively removed with mercuric chloride. Palo (ref. 79) employed the same technique for qualitative headspace analysis.

Little attention has been given to the design of the headspace equilibration equipment. Reliable performance of the equipment is one of the most important requirements for quantitative headspace analysis. One must be able to control the operating conditions and be able to reproduce them. It is more important for headspace gas analysis than for other quantitative techniques. The chemist must be careful in both collecting and transferring the sample so that changes do not occur because of condensation or leakage. Clean equipment is essential so that residual components are not present from a previous sample. This is especially important in environmental studies because of the concern about concentration levels.

Calibration of such a system can be accomplished by separate preparation of an appropriate model standard, by direct addition of a standard to the sample after initial analysis, or by prior determination of the distribution constant with a defined liquid phase composition. An aliquot of the equilibrated gas phase is transferred to a gas chromatograph for analysis using either a gas-tight syringe or a gas-sampling valve (in certain circumstances). The method can be applied to a wide variety of volatile organic compounds which are pollutants in water. In every case investigated (with normally volatile compounds), the application of this technique is usually sensitive to levels of 10 ppb or less in the original sample.

A variant of the headspace gas analysis technique has been described by Gottauf (ref. 80) in which he determined the trace volatiles in water samples. The headspace gases were concentrated by adsorption on a cooled adsorbent and losses by adsorption or adsorption in the original sample were minimized. The limit of detection was below 10^{-8}% with an R.S.D. of $\pm$0.05%.

The type of material used in the seal when sampling by means of a syringe has been shown to introduce errors. Maier (ref. 81), Davis (ref. 82) and Gilliver and Nursten (ref. 83) independently have demonstrated the loss of sample components because of rubber seals (e.g., serum bottles). Similar losses were noted with silicone rubber septa. Davis's (ref. 82) data revealed that exposure to rubber seals for thirty minutes decreases component concentration of hydrocarbons and aldehydes.

The rate of diffusion among the various components is another source of error for quantitative data by this technique. This error becomes pronounced when the equilibration time is short and the headspace volume is large (ref. 84). The advantages and disadvantages of sampling and injecting with a syringe have been treated by Binder (ref. 84).

The initial result of the headspace gas analysis is only a relative measure because the matrix gas is not measured. If the gas contains more than one component (which it usually does), the calculation of the concentration for the nth component may not necessarily be valid because a linear relationship will not exist between partial pressures and concentrations unless the solutions and gaseous phases both approach ideality. Since most headspace gases are mixtures of several volatile components, the total pressure (vapor pressure) is the sum of the partial vapor pressures:

$$P_t = P_1 + P_2 + P_3 + \ldots + P_n = x_1P_1^o + x_2P_z^o + \ldots + x_nP_n^o \quad (4.10)$$

where P_t = total vapor pressure of system;
P_1 etc. = partial pressure of each component;
P_1^o etc. = vapor pressures of the pure components;
x_1 etc. = mole fractions of each component.

Using eqn. 4.10, one assumes that the partial pressures are sufficiently small so that they may be identified with Dalton's partial pressures:

$$P_n = (n_nRT)/V \quad (4.11)$$

This implies that the number of moles n_1, n_2, ...n_n are easily determined from the chemical analysis of the vapor volume, V. Thus, one must consider two situations:

(1) As the pressure of the headspace gas increases, complications arise because of deviations from the ideal gas laws.

(2) Total pressure (P_t) dependence on the sum of the partial pressures varies from system to system, and it is more accurate to incorporate activity coefficients γ in eqn. 4.10.

$$P_t = \gamma_1x_1P_1^o + \gamma_2x_2P_2{}^o + \ldots\ldots + \gamma_nx_nP_n{}^o \quad (4.12)$$

Because of the variety of samples encountered in headspace gas analysis, a number of situations may present themselves.

(A) The components of the sample could show no mutual interaction (i.e., similar type molecules). Linear dependence of the total pressure with the partial pressures will follow, as shown in eqn. 4.10. In this case, the activity coefficients γ, will equal 1.

(B) The interactions between like molecules are greater than the interactions between unlike molecules in the vapor phase. The like molecules tend to displace each other from the mixture and increase their partial pressures. This results in a positive deviation from Raoult's law. It also means a liquid phase of these components will boil at a temperature lower than either of the pure liquids. This occurs when some of the molecules have polar groups and some of the molecules have non-polar groups (e.g., alcohols and hydrocarbons). This is predicted by eqn. 4.12, when the activity coefficients, γ, are not equal to 1.

(C) The reverse of situation B could occur (i.e., the interaction between unlike molecules is greater than the interaction between like molecules). This decreases the partial vapor pressures from the values for an ideal mixture and causes negative deviations from Raoult's law. The liquid phase of the mixtures of these components will boil at a temperature higher than either of the pure components. Systems like this occur when the attractive forces are of electrostatic origin, i.e., the molecules possess permanent or induced dipoles which can lead to the formation of hydrogen bonds or other labile bonds between the molecules.

(D) A combination of B and C, i.e., both positive and negative deviations in the gas phase.

Situations B, C and D will, in many cases, show partial pressures not varying linearly with concentration or which are completely independent of concentration in the high-concentration ranges. Headspace gas analysis under these conditions will not give quantitative results or may even give the same result for different concentrations. The systems described in B, C and D can be minimized if the experimental work is performed at high dilution (region of ideal mixtures). Hachenberg (ref. 85) has shown that acrylonitrile may be determined in water at low concentrations using headspace gas analysis; in fact, the relationship between peak area and concentration is linear. Ozeris and Bassette (ref. 86) demonstrated that, at low concentrations, ketones, aldehydes, alcohols, diethyl sulfide and esters show a linear relationship between peak height and concentration (0.01-10 ppm levels in water).

Sometimes the concentrations of the components of interest are at levels where the detector signal is of the same order of magnitude as the noise. If this occurs, one simply enriches the components in the headspace gas. The easiest way to enrich the headspace gas is to increase the ionic strength of the aqueous solution. The addition of highly soluble salts will do this very conveniently. Salts such as sodium sulfate (ref. 87), ammonium sulfate or sodium chloride (ref. 87) have been used. Even calcium carbonate (which is not that soluble in water) has been used successfully (ref. 88).

The big advantage of headspace gas analysis by gas chromatography is that the solvent peak is eliminated. Combining this advantage with the enrichment technique permits one to do an analysis with greatly reduced detection limits. This advantage applies equally to systems where one is dealing with nonaqueous solvents (ref. 88).

The distribution contant of a solute in headspace gas analysis is really the inverse of Henry's law; however, the chemist is more concerned with the mass balance of the solute between the two phases. Thus, he finds it more convenient to define the distribution constant in terms of the ratio of the equilibrium concentrations of the solute in the condensed and gaseous phases.

$$K_D = (W_L/V_L)/(W_G/V_G) \tag{4.13}$$

where W_L and W_G = mass of the solute in the condensed and gaseous phases, respectively, and V_L and V_G = volumes of these two phases.

It is fairly easy to obtain quantitative data if one has a one-phase system. However, in headspace gas analysis the solute component is distributed between two phases. Thus one must know the magnitude of the mass distribution between the two phases. To obtain this information all one needs is the total amount of the solute in the system and the amount of the solute which has been equilibrated into the gas phase. Thus the total amount of the solute, W_n, can be expressed as:

$$W_n = c_{nG}(K_DV_L + V_G) \tag{4.14}$$

where $c_{nG} = W_{nG}/V_G$.

By quantitative gas chromatography one can obtain a value for c_{nG}. The values of V_L and V_G are known and if one knows K_D, the value of W_n is easily calculated. The problem is that one very seldom knows the value of K_D. This problem may be circumvented in several ways (ref. 56).

(1) Use experimental conditions in which the value of K_D is essentially equal to zero. This may be achieved by working at elevated temperature. In this case, eqn. 4.14 becomes:

$$W_n = c_{nG}V_G. \tag{4.15}$$

Using quantitative gas chromatography, it is easy to determine the value of c_{nG}. The method is very seldom used; preference is given to ambient conditions.

(2) Use a standard system solution, i.e., a sample which contains <u>all</u> the components of the unknown sample <u>plus</u> a known amount of the sample component being determined. If one keeps V_L and V_G the same for both samples, the K_D for both samples has to be the same. This being the case, the factor $(K_DV_L + V_G)$ will be eliminated because it is common to both systems. If W_n* is the total amount of the solute

component in the standard system solution and c_{nG}* is its concentration in its gas phase one obtains:

$$W_n^* = c_{nG}^*(K_D V_L + V_G) \quad (4.14A)$$

and

$$W_n = c_{nG}(K_D V_L + V_G). \quad (4.14B)$$

Dividing eqn. 4.14B by eqn. 4.14A:

$$W_n = W_n^* c_{nG}/c_{nG}^*. \quad (4.16)$$

Determining c_{nG} and c_{nG}* under the same gas chromatographic conditions gives us the value for W_n. This method is not used because it is almost impossible to make a standard system solution which is the same as the unknown.

(3) Use of the standard addition method. The assumptions made here are that the added amount of the component to be determined involves little or no change in the values of V_G and V_L (i.e., the ratio of V_L/V_G). This, in turn, causes little or no change of the thermodynamic properties of the solution and therefore K_D does not change. The main constraint placed on this method is that the sample, before and after the addition of the known amount of component n, must by analyzed under identical conditions. The amount of added component must be greater than the estimated concentration in the unknown (a minimum of 2-3 times greater).

Following these conditions the quantity W_n may be calculated:

$$W_n = (W_s - W°_n)/((c°_{nG}/c_{nG}) - 1) \quad (4.17)$$

where W_s = mass of added standard;
$W°_n$ = estimated amount of component n before standard addition;
$c°_{nG}$ = concentration in gas phase after standard addition.

This third method of analysis is the one recommended for the analysis of aqueous waste effluents for the presence of priority pollutants.

There are a number of good publications concerning headspace analysis: a review written by Drozd and Novak (ref. 56), two books (refs. 89 and 90), and a recent paper describing its applications to the analysis of priority pollutants in waste water effluents (ref. 91). Any of these references will furnish the needed background for someone using the technique for the first time.

McAuliffe (ref. 35) published a method of headspace gas analysis which is more specifically titled "vapor equilibration analysis". His original work was concerned with hydrocarbons in water but has now been extended to many other types of compounds. The technique involves equilibrating equal volumes of an aqueous solution and an inert gas (usually helium or nitrogen) in a large glass syringe. At this point in the procedure the chemist has two options:

(1) The entire gas phase may be injected into the sample loop of a gas sampling valve mounted on the gas chromatograph. A second equal volume of inert gas is drawn into the syringe. The procedure is repeated several times. The original concentration of components in the sample can be calculated from the dependence of concentration in the headspace on the serial number of extractions performed.

(2) After filling the syringe with equal volumes of aqueous sample and inert gas (clean air may also be used), the syringe may be capped with a rubber septum and aliquot volumes removed by means of another smaller syringe. The content of this smaller-volume syringe may be injected into the chromatograph through the septum of the sample injection port. Replicate samples may be injected without seriously affecting the quantitative results of the method.

The method of standard addition may also be used with this syringe vapor equilibration technique. This has been demonstrated repeatedly (refs. 64, 92 and 93). One removes a volume V_G of the headspace gas from the equilibrated system and injects it into the gas chromatograph. The component of interest will have a peak area A_n. A known mass W_s is then introduced into the equilibrated syringe system and allowed to re-equilibrate. A second volume $V°_G$ is then removed and injected into the gas chromatograph giving a peak area $A°_n$. The mass of the component in the original sample may then be calculated:

$$W_n = (W_s - W°_n)/((A°_n V_G/A_n V°_G) - 1) \qquad (4.18)$$

Using this technique, a hydrophilic solute (1-100 ppm) and aliphatic and aromatic hydrocarbons (0.001 - 20 ppm) have been determined with relative errors of approximately 20-22% and 10-12%, respectively (ref. 94).

Currently the method developed by Bellar and Lichtenberg (refs. 95-95) has been widely applied for similar analyses. Their method, commonly designated "purge and trap", differs from the vapor-equilibration method. The purge and trap method requires significantly more elaborate equipment and a reliance upon either 100% trapping efficiency of each component of concern or a careful delineation of the trapping efficiency and precise duplication of sample processing. Equipment cost for vapor-equilibration analysis is far lower, and its use is considerably simpler. In a number of cases the purge and trap system may be more sensitive because the total volatile organic content of an aqueous sample (5-60 ml) can be transferred to the instrument for a single analysis. This is generally not true in the case of vapor equilibration analysis. However, replicate analyses of the same sample are much more readily obtained by vapor equilibration. The purge and trap system requires a transfer of the sample from its original container to the purge system. Unless extreme care is exercised with this step, the most volatile and/or least soluble of the organic contaminants can be lost in unknown proportion. Alternatively, even if the purge system container is used to obtain the sample (not common practice because of the limited sample size), it must generally be opened in order to mount it in the instrument system, with potential attendant vapor losses.

One manufacturer (refs. 98-101) uses what is designated as the "dynamic technique of analysis." Volatile organic components in a water sample are continuously purged by a gas and then quantitatively transferred to an adsorption trap where they are later thermally desorbed and backflushed onto an analytical gas chromatographic column. Another manufacturer (refs. 102-105) has offered a dedicated, multi-sample unit in which the detectors and control units are modular. Thirty samples can be accommodated with fully automatic operation. The unit offers a choice of flame-ionization detector, phosphorus detector, electron-capture detector or hot-wire detector, in addition to column-effluent splitting and dual-detector operation. The unit will accept liquid or solid samples and can be used for alcohol in blood, residual monomers in polymers, industrial hygiene (e.g., vinyl chloride in air), forensic work (amphetamines in urine) and environmental studies (e.g., hydrocarbons in water).

With vapor-equilibration analysis it is possible to take the sample in the same container ultimately used for its analysis. This container is septum-sealed so that, once obtained, the sample is never again opened, even during the analysis. Vapor losses are precluded as long as the septum material is not permeable to the sample components of interest. A number of such septum materials are available. The vapor equilibration analysis method has shown distinctive superiority because of this simplified sampling in the analysis of chemical process waste lines which are obtained at elevated temperatures.

Residual vinyl chloride in poly(vinyl chloride)(PVC) may be extracted by acetone and injected directly into the gas chromatograph or a sample of PVC may be heated (sealed container) with a small amount of diethyl ether and a headspace sample taken (ref. 106). Both techniques are considered good approaches to the problem. Cowen et al. (ref. 107) reported the construction of a device to aid in septumless injections of headspace gas and compared it to the syringe injection technique.

In a typical analysis, using McAuliffe's technique (ref. 35), one would use a 50-ml glass syringe containing 25 ml of the water sample and 25 ml of an inert gas. After the two phases have equilibrated, one may use the entire gas phase or an aliquot taken with a smaller syringe (see above). Unless one knows the value of the distribution constant K_D, the data will be more qualitative than quantitative. Loss of any water phase is not a problem; the second gas equilibration is carried through with a gas volume equal to the decreased liquid volume. The critical parameter is control of temperature; this should remain constant from one equilibration to the next.

Determination of the distribution constant K_D can easily be performed.

(A) Equilibrate equal volumes of inert gas and water sample in a 50-ml syringe.

(B) Inject a known volume of the gas phase into the gas chromatograph and measure either the peak area or the peak height and relate this measurement to concentration.

(C) Remove any remaining gas phase in the 50-ml syringe.

(D) Equilibrate the water phase with several additional equal volumes of inert gas and repeat steps B and C with each.

(E) It can be shown (ref. 35) that the following equation will relate K_D to concentration in the water phase:

$$\log ((X)_g)_n = -n\log(K_D + 1) + \log K_D(X)_1 \tag{4.19}$$

where $((X)_g)_n$ = quantity of component X in the gas phase after n equilibrations, and $(X)_1$ = initial quantity of X in the water phase.

A log plot of area or peak height (in terms of concentration) versus the number of equilibrations will produce a straight line having a slope of $-\log(K_D + 1)$, and an intercept of $\log K_D(X)_1$. A representation of such a plot is shown in Fig. 4.2. Having this plot, one can select any two adjacent gas phase concentration points, divide the smaller value into the larger value and subtract one from the quotient to obtain the numerical value of $\log K_D$. One then divides the intercept $\log K_D(X)_1$ by

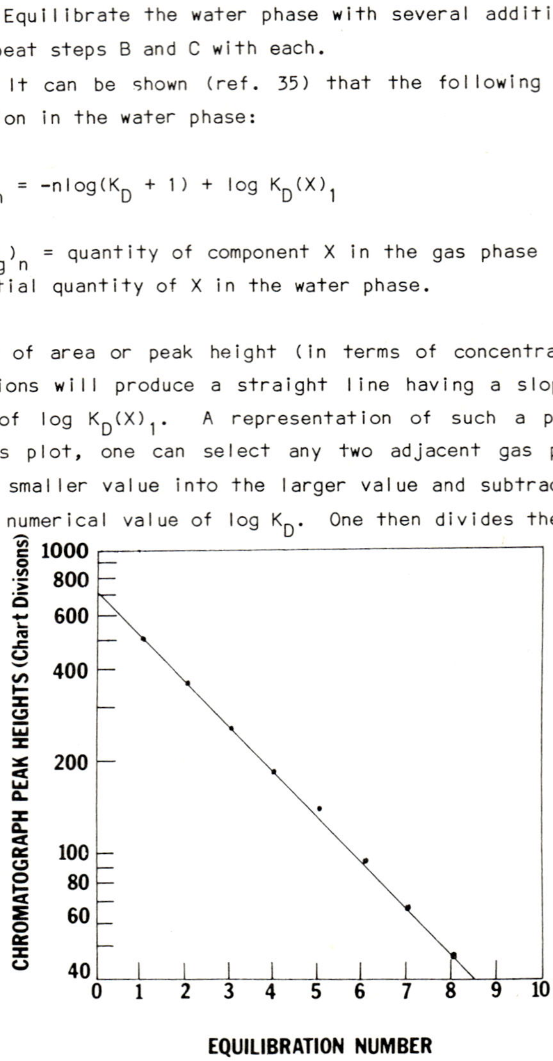

Fig. 4.2 Partitioning of a solute between equal volumes of H_2O and an inert gas (e.g., helium)

log K_D and obtains $(X)_1$, the concentration of the component of interest in the original water sample.

In addition to the ease of obtaining the numerical value of K_D, as described above, this technique has two other outstanding features.

(1) Most environmental water samples need only be equilibrated two times. The final data will not be as accurate as a multi-point plot but will have a high degree of accuracy. If two equilibrations are performed, divide the concentration from the first equilibration $((X)_g)_1$ by concentration from second equilibration $((X)_g)_2$. Multiply this quotient, Q, by the first concentration to obtain the intercept value. Knowing the K_D, (previous analysis of samples from this system) one divides the intercept by K_D to obtain the concentration of the component in the water sample.

$$((X)_g)_1/((X)_g)_2 = Q \tag{4.20}$$

$$Q \times ((X)_g)_1 = K_D(X)_1 \tag{4.21}$$

$$K_D(X)_1/K_D = (X)_1 \tag{4.22}$$

(2) If the water samples come from a given location, which is being monitored, (i.e., if one has a "constant" matrix composition), the K_D value need be checked periodically (it should be constant) and each sample can be equilibrated only once. One may then calculate the concentration of X in the initial sample by:

$$((X)_g)_1((1 + K_D)/K_D) = (X)_1 \tag{4.23}$$

The quantity of organic material found in the vapor phase is dependent upon four variables: (a) concentration; (b) partial pressure; (c) temperature; and (d) ionic strength. McAuliffe (ref. 108) presented the results of the effects of these variables at a symposium on environmental analysis. The results of his study showed: (1) a decrease in solubility (increase in quantity found in gas phase) of six to seven orders of magnitude as the n-alkane is changed from methane to dodecane; (2) aromatic hydrocarbons (C_6 - C_{24}) showed a similar decrease in solubility or increase in partial pressure; and (3) aromatic hydrocarbons are more soluble than n-alkanes or cycloalkanes with the same number of atoms (e.g., benzene > cyclohexane > hexane).

The method of vapor equilibration has been shown to have high precision (ref. 35) provided one controls the gas:water volumes and temperature (both remain constant throughout replicate measurements). Using hexane, cyclohexane and toluene as standards, McAuliffe obtained standard deviation of $\pm 0.7\%$, $\pm 0.8\%$ and $\pm 0.5\%$, respectively.

This method is not restricted to systems where the gas:water (sample) ratio is 1:1. It has been shown (ref. 108) that an increase in the gas:water ratio gives a corresponding percentage increase in transfer of the individual hydrocarbons to the gas phase. Alkanes, alkenes and cycloalkanes show a decrease in concentration per unit volume of gas phase whereas aromatic hydrocarbons give essentially a constant concentration per unit volume of gas phase over the gas:water ratio ranges of 1:10 and 10:1. One final comment: Although the hydrocarbon classes partition to almost the same extent, they all demonstrate an increase in partitioning to the gas phase with an increase in molecular weight. One would expect this trend because water solutilities (Δ solubilities) decrease more rapidly than vapor pressures increase (ΔP_i).

4.3.2.1.1 Preparation of gas standard mixtures. Let us describe the preparation of a gas standard for these types of analyses. It is the authors' opinions that laboratory-prepared standards are easy to prepare and thus preferable to other options where care is properly exercised. It is for this reason that the details are presented here. One should use appropriate safety precautions for handling compressed gases and cylinders (Chapter 7). For 1000 ppm standard gas mixture, one takes 100 ml of an inert gas (helium or nitrogen) in a standard 100-ml syringe which also should contain some 2-mm glass beads (6-8 beads) for mixing. This 100-ml syringe is capped immediately with a red rubber septum cap over a needle hub. A lecture bottle containing the pure standard gas(es) has either a commercial pressure-reducing valve system or a standard lecture-bottle valve added. The lecture-bottle system with the added valve or pressure reducer is fitted with a septum seal at the exit end. All the valve system(s) between the exit and the lecture-bottle valve are opened and the enclosed volume is evacuated by means of a large standard glass syringe by inserting the syringe needle (with the syringe plunger all the way down in the syringe barrel). The plunger is pulled back to the limit either of internal volume or of the operator's ability to withstand the atmospheric pressure opposition. If the syringe chosen for this purpose is too small, this procedure should be repeated until an evident vacuum is contained in the interior space of concern. The auxiliary valve is then closed and the main tank valve is opened long enough to fill the space between the two valves and then immediately closed. Insert a standard glass syringe carefully through the mounted septum and then open the auxiliary valve. The limiting of the volume segments should allow a 5- to 10-ml syringe to absorb all the gas expansion which results without exceeding either the syringe capacity or the operator's ability to control the syringe plunger. Close the auxiliary valve, withdraw the syringe and its contents and discharge to an adequately ventilated system. Any residual pressure in the space between the auxiliary valve and the septum should be relieved by again inserting the syringe and withdrawing gas until an evident vacuum exists within the active space. This entire process should be repeated at least three times to minimize the dilution of pure gas with air which was originally contained in the system and to avoid bleeding unreasonable quantities of potentially toxic gas into the laboratory atmosphere. Finally, the last sequence involving opening of the second valve should be done with an empty syringe needle through the septum to act as a final purge. Following this, the empty needle should be withdrawn and the auxiliary valve closed. The last step involves first, opening the main tank valve and closing it; second, opening the auxiliary valve and closing it, removing 0.2 to 0.3 ml of the contained gas into a lockable valve syringe such as the Pressure-Lok Series A-2 syringe, followed by closing the syringe valve. The removed volume should be discharged to the ventilation system with opening of the syringe valve which should be immediately closed. The syringe is again inserted through the septum and 0.2-0.3 ml, transferred to the syringe, and the syringe valve closed. The syringe plunger should be collapsed to the 0.1-ml mark and excess pressure relieved to this point. To assure that an excess pressure exists, the needle tip should be placed below a liquid surface to allow for visual observation of the release of the gas. The syringe valve should be immediately closed after relieving the pressure. The needle of this syringe then should be placed through the syringe cap on the 100-ml syringe and the 0.1-ml volume transferred to the larger syringe. If gas mixtures are desired, this process must be repeated with each individual gas. Lower error will result if one prepares a 1000-ppm standard mixture (1-ml pure standard gas + 99-ml of matrix gas) and subsequently removes 10-ml of the carefully mixed 1% standard and transfers this to a second 100-ml syringe containing 90-ml of clean matrix gas.

4.3.2.1.2 Error involved when preparing a binary gas mixture. One may calculate the ppm of a standard gas in a diluent gas (matrix gas) by the following equation:

$$C_{ppm} = \frac{V_s}{V_D + V_s} \times 10^6 \qquad (4.24)$$

where V_s = volume of standard gas and V_D = volume of diluent (matrix) gas. The V_s term in the denominator may be neglected when preparing concentrations $\leq$ 5000 ppm because the contribution of V_s to total volume is insignificant. Table 4.3 lists some representative data for the preparation of gas standards. Examination of the data in Table 4.3 demonstrates the fact that any error involved in using a 100-ml syringe for diluent (matrix) gas and a 1-5-ml syringe for the solute standard gas is very small.

4.3.2.2 Purging, sparging and/or vapor stripping. If the organic components of a water sample are volatile or can be volatilized, one may analyze the resulting vapor with a high degree of sensitivity. The three terms purging, sparging and vapor stripping essentially mean the same thing, i.e., a technique wherein an inert gas is used to evacuate or free organic components by means of agitation of the sample. What is different among the three techniques is how it is performed and how the equipment is physically set-up. Purging of a water sample usually implies a continuous flow of

TABLE 4.3

Contribution of V_s term in denominator of Eqn. 4.24

Accounting for $V_D + V_s$ term (ppm)	Neglecting V_s term (ppm)	Relative error (%)
1	1.000001	10^{-4}
10	10.0001	10^{-3}
100	100.01	10^{-2}
1000	1001.0	10^{-1}
5000	5025.0	0.5

Relative error = ((exact value - approximate value)/exact value) x 100.

an inert gas through the sample and the passing of the volatiles through a trapping tube wherein the volatile components are adsorbed. Sparging of a solution implies agitating the solution with an inert gas by means of a sparger (e.g., a fritted disc) and collecting volatiles either by adsorption on a solid adsorbent (Tenax GC or charcoal) or freezing-out the volatiles in a trap submerged in dry ice acetone, liquid nitrogen, etc. Vapor stripping is an all-inclusive term which could refer to either technique.

If the volatiles are adsorbed on a solid, they may be desorbed and backflushed into a gas chromatograph. If the volatiles have been condensed in a cooled trap, the final sample will contain a certain quantity of water, which may interfere with the gas chromatographic analysis. This problem may be minimized by preceding the cooled trap with a desiccant tube. The presence of the desiccant tube eliminates one problem but creates another, i.e., some of the volatiles are adsorbed on the desiccant.

The technique of purging a solution of its volatile components and trapping them is commonly referred to as the "purge and trap" method of separation. The method is credited to Bellar and Lichtenberg (refs. 95-97 and 109). The U.S. EPA has adapted this separation technique to several of its analytical procedures (refs. 110-112).

The technique of sparging a water sample of its volatiles and collecting them in a cooled trap is credited to Swinnerton and Linnenbom (refs. 38, 42 and 113-115). Their main work in this area was the identification and quantification of lower hydrocarbons in seawater. A sensitivity of 0.1 ppt was achievable because of the large sample volume and the concentrating step in the procedure.

Vapor stripping techniques present a problem of calibration because the efficiency is a function of the solute and the ionic strength. The use of an internal standard or an internal standard of a labeled isotope of the component of interest (e.g., deuterated benzene) will eliminate this problem (ref. 116).

This sample treatment technique is very similar to the sampling technique by adsorption discussed in Chapter 3 (Section 3.4). The reader is referred to this chapter for more information.

Several boiling point limits (for purgeable compounds) have been stated in the literature. Grote and Westendorf (ref. 117) set the boiling point limit at 150°C and a water solubility of 3%, whereas Hites (ref. 118) has stated a limit of 100°C for the boiling point.

Bellar et al. (ref. 119) applied their "purge and trap" technique to the determination of vinyl chloride, at the μg/ℓ level by gas chromatography. Nitrogen was the purging gas and the adsorbent was either silica gel or Carbosieve at room temperature. A halogen-specific detector was used. Some new columns were used for the separation and determination of organic pollutants in water by Mindrup (ref. 120). The volatiles were removed by "purge and trap" technique and determined on a 80-100 mesh Carbopack C/0.2% Carbowax 1500 column. Dowty et al. (ref. 121) determined halogen and nonhalogen containing priority pollutants at the < 1 μg/ℓ (1 ppb) level. They increased their sensitivity by use of "purge and trap" and mass spectrometry. Moye (ref. 122) trapped polar water-soluble compounds on Tenax GC after gas-stripping from the sample. The types of compounds tested were alcohols, aldehydes, ketones, ethers, halogenated hydrocarbons and aromatic hydrocarbons. With the exception of methanol and ethanol, most of the strippings, adsorptions and desorptions were at least 75% efficient. Dissolved nitrous oxide in seawater was stripped (purged) using nitrogen gas. The gas was adsorbed on molecular sieve 13X at 0°C. Thermal desorption was done at 250°C and the nitrous oxide determined on a molecular sieve 5A column at 250°C using electron-capture detection (ref. 123).

4.3.2.3 Change in ionic strength of sample solution. The removal of nonpolar or compounds of low polarity from the aqueous phase into the gas phase may have to be enhanced for several reasons.

(1) The concentration level at which they are present is very low and vapor equilibration, purging or sparging may not transfer a sufficient amount to the gas phase for accurate detection.

(2) One desires to approach quantitative transfer (at least 95%) of the components from aqueous phase to gas phase.

The simpliest technique to increase the amount of pollutant in the gas phase is to increase the ionic strength of the sample. The result is that the low polarity compounds will have more interaction among themselves which results in an increase in their partial pressures. Thus, one has "squeezed" these low polarity molecules out of the aqueous phase (i.e., one has salted them out). This is accomplished by adding enough of a soluble electrolyte to form a very concentrated solution or in some cases to saturate the aqueous phase. Compounds such as NaCl, Na_2SO_4 and $(NH_4)_2SO_4$ (ref. 87) as well as $CaCO_3$ (ref. 88) have been used.

The same effect is seen in performing a liquid-liquid extraction with water as one of the phases. The addition of a large salt concentration to the water phase would enhance the extraction of low polarity compounds into the organic layer. It is not within the scope of this book to discuss the theory of ionic strength, salting out, etc. One may find information regarding its benefits by reading the appropriate section in a physical chemistry textbook. An excellent treatment of this topic is given by Glasstone (ref. 124). Drozd and Novåk (ref. 92) have studied the effect of increased ionic strength for the headspace gas analysis technique, by standard addition, for hydrophilic solutes (e.g., acetone, methanol, ethanol and propanol). The increase in ionic strength demonstrated an increased sensitivity for acetone and propanol.

4.4 CLEAN-UP

Sample clean-up is used when an environmental sample contains or is suspected to contain a wide variety of components. The goal of this procedure is to isolate groups of compounds so that injection of an aliquot of the extracting solvent does not result in a very complicated chromatogram (i.e., the removal of unwanted components which leave the data in doubt as to reliability). The simpliest approach to this problem is to pass the sample as received (or an extract of the sample) through a clean-up column to obtain an initial group separation.

Activated charcoal (refs. 125 and 126) and the various porous organic polymers are widely used to clean up and concentrate pollutants from water. The two techniques, carbon-chloroform extract (CCE) and carbon-alcohol extract (CAE) are two illustrations of adsorption-extraction systems for clean up of water samples. Charcoal has advantages (e.g., high adsorption capacity and high thermal stability) and disadvantages (e.g., organic compounds are not adsorbed completely; desorption is not always quantitative and the desorption process sometimes changes the sorbate molecules) (ref. 127).

Porous polymers are good for cleaning up water samples because the proper sorbent causes a shift in equilibrium towards the sorbent more than toward the organic phase in a liquid-liquid extraction. The concentration factor will be greater for liquid-solid extraction than for liquid-liquid extraction. Porous polymers have some unique properties which make them the choice for cleaning up and concentrating water pollutants. If the proper porous polymer is selected, the distribution constants (K_D) tend toward infinity. Adsorption of water is minimal at best; the porous polymer surface is inert; and their "wetability" is low which increases the adsorption of organic pollutants from the water. The more hydrophobic is the sorbent surface, the more adsorption capacity is available for non-polar (or low polarity) molecules in the water phase (as molecule polarity decreases, the water solubility decreases).

Removal of organic molecules from the surface of the porous polymer may be accomplished by liquid-desorption (see above, e.g., CCE and CAE) or by heating. Solvents other than chloroform and alcohol which have been used successfully are diethyl ether, n-hexane, acetone, isopropanol, methanol, methyl isobutyl ketone and pyridine. Experience with thin layer or liquid-solid chromatography will be advantageous in the sample clean-up process and familiarity with Snyder's (ref. 128) eluotropic solvent series would be a great asset to the novice. A word of caution in the use of the eluotropic solvent series: when using non-polar porous polymers an increase in solvent strength will give the opposite effect (more adsorption and less desorption). The best solvents for desorption would follow the order benzene > n-hexane > ethyl acetate > butanol > diethyl ether > propanol > acetone > ethanol > methanol > water (ref. 10).

Desorption volume of the eluent is usually in the range of 10-25 ml. This liquid desorption step as well as the concentrating step (evaporation to a smaller volume) which follows, introduces error. The large amount of eluent in the final sample may also cause an error because of the large peak which results from the solvent. It may interfere or completely cover a pollutant peak. Sometimes the solvent peak may be removed, e.g., if pyridine were used for the desorption, the sample pulse could pass through a small precolumn containing $CuCl_2$ and the pyridine selectively removed (ref. 129). NOTE: <u>The solvent(s) used for the desorption must be of high purity - they may contain no more than a ppm or ppb of the component one is measuring</u>. One should purify all solvents used, unless, when they are chromatographed, at a high sensitivity setting, no interference peaks are noted.

Table 4.4 is a tabulation of a number of commonly used solids in clean-up and concentration of pollutants in water. We have given properties and selectivities of these various solid adsorbents so that they may be used as a reference point in water pollution analysis. Acidic compounds were found to be easily desorbed with aqueous NaOH, whereas basic components are eluted with dilute HCl.

TABLE 4.4

Solid sorbents used for clean-up and concentration of water pollutants

Sorbent	Chemical make-up	Chemical and physical properties	Compounds sorbed*	Recovery (%)	Ref.
Amberlite XAD-2	Styrene-divinyl-benzene copolymer	Hydrophobic. Pore diameter, 85-90 A. Specific surface area: 290-330 m^2/g. 20-50 mesh.	Alcohols (10-100 ppb)	91-102	130
			Aldehydes and ketones (10-100 ppb)	92-102	130

Sorbent	Chemical make-up	Chemical and physical properties	Compounds sorbed*	Recovery (%)	Ref.
			Esters (10-100 ppb)	91-103	130
			PAHs (10-100 ppb)	92-101	130
			Alkyl-benzenes (10-100 ppb)	ca. 93	130
			Halogenated aromatic hydrocarbons (10-100 ppb)	93-99	130
			Aromatic amines (10-100 ppb)	91-100	130
			Nitro-compounds, anilines, quinolines, (10-100 ppb)		
			Ethers (aromatic) (10-100 ppb)	91-99	130
			Pesticides herbicides (20 ppt)	93-96	130
Amberlite	Styrene-divinylbenzene copolymer	Hydrophobic. Specific surface area: 750 m^2/g.Av. pore diameter, 50 A.	Alcohols (2-10 ppb)	93-103	131
			Esters (2-10 ppb)	91-99	131
			Aldehydes and ketones (2-10 ppb)	92-105	131
			Alkyl benzenes (2-10 ppb)	(65-80)	131
			PAHs (2-10 ppb)	(83-87)	131
			Phenols (alkylated) (2-10 ppb)	93-105	131

Sorbent	Chemical make-up	Chemical and physical properties	Compounds sorbed*	Recovery (%)	Ref.
			Aromatic chloro compounds (2-10 ppb)	91-93	131
			Acids (2-10 ppb)	90-107	131
			Phenols (halogenated) (2-10 ppb)	91-99	131
Amberlite XAD-1	Styrene-divinylbenzene copolymer	Specific surface area: $100 m^2/g$. Mean pore diameter: 200 Å.	Detergents, pH = 2.0	100	132
			Insecticides, pH = 2.0	100	132
			Methylene blue, rhodamine blue, pH = 7.6	100	132
			Humic acids. pH = 2.0	100	132
			Vitamins B_2 and B_{12} pH = 2.0 and 7.6 respectively.	100	132
			Cholesterol pH = 2.0	100	132
Chromosorb 102	Styrene-divinylbenzene copolymer	Hydrophobic. Specific surface area 300-400 m^2/g. Mean pore diameter: 95 Å.	Chloroform	93	133
			Benzene	100	133
			Methyl isobutyl ketone	100	133
			o-Ethyl phenol	97	133
Tenax-GC	Poly (2,6-diphenyl-	Specific surface area: 19-30 m^2/g. Pore volume: 0.667	Pesticides (free Cl_2 in water	90	135

Sorbent	Chemical make-up	Chemical and physical properties	Compounds sorbed*	Recovery (%)	Ref.
	p-phenylene oxide)	ml/g (ref. 134). Av pore radius: 720 A. Good sorbent up to 320°C. Reaction with nitrogen oxide or nitric acid does not affect its efficiency or capacity.	causes oxidation of pesticides)		
			PAHs (0.1 ppb level)	85-90	135
			Benzene, aniline, p-cresol.	89-99	136
Polyurethane, open pore	Urethane polymer	1-10 µm agglomerated spherical particles bonded together. Bases break up open-pore structure. Stability increased as pH of water sample decreased.	PAHs	92-108	137
Polyurethane porous foam	Urethane polymer.	pH 6-9. Increased water flow rate causes decreased adsorption.	Chlorinated insecticides and poly-chlorinated biphenyls.	97-114	138
			PAHs	pH 3.0,62 pH 10.0,76	139
Amberlyst A-26 ion exchange resin.	Macroreticular porous polymer.	Anion-exchange resin with trimethylamine group. Specific surface area: 25-30m^2/g. 27% porosity.	Phenols (pH 12-12.5)	95-102	127

*By group or individual compound listing.

**Individual compounds listed had to have a minimum of 90% recovery. Recovery (%) = ((Amount remaining after concentration step)/(Amount originally present in solution)) x 100.

Once the components of the water sample have been removed (desorbed) a group separation may be used to selectively remove groups of compounds. This group separation is dependent on pH (ref. 140) because strongly ionic organic or inorganic compounds are not sorbed.

WATER SAMPLE clean-up by adsorption → COMPONENTS CONCENTRATED ON SOLID SORBENT
↓
Wash with 0.05 M NaOH → REMOVAL OF ACIDIC SORBATES → Wash with 1 M HCl*
↓
REMOVAL OF BASIC SORBATES → Wash with diethyl ether → REMOVAL OF NEUTRAL SORBATES

* Recoveries 7-75 % with only distilled water; 40-108 % with HCl.

Use of a finer particle size (ca. 150 mesh), varying solute concentration from ppb to ppm range, changing the flow rate of the mobile phase, and the addition of up to 50 g NaCl per liter produced little or no change in the recoveries.

The Amberlite XAD-7 sorbent is a hydrophilic methacrylate polymer having a specific surface area of 450 m^2/g and a mean pore diameter of 80 Å. This is a good clean-up sorbent for fatty acids, phenol, and m-chlorophenol. Adsorption increases as the fatty acid chain length increases (solubility decreases). The polymer is unstable in alcoholic sodium hydroxide solution (causes hydrolysis of the ester group).

Amberlite XAD-8 resin is a methyl methacrylate resin with average pore diameter of 250 A and specific surface area of 140 m^2/g. The resin adsorbs aliphatic systems in preference to alicyclic and/or aromatic carbon systems. The preference for functional groups is $-CH_3 > -COOH > -CHO > -OH \geq -NH_2$. Both of these preferences are an inverse of solubility trends. One of its main uses is the concentration of low molecular weight organic solutes and natural organic polyeloctrolytes (e.g., fulvic and humic acids).

Another group of sorbents which may be used for clean-up is the Porapak Polymers (manufactured by Waters Associates, Framingham, MA, USA). Porapak Q is an ethylvinylbenzene-divinylbenzene copolymer having a mean micropore diameter of 74.8 Å available up to 500 Å) and a specific surface area of 630-840 m^2/g (refs. 141 and 142). The polymer has its best application for the concentration of aromatic hydrocarbons (e.g., benzene, toluene, o-xylene, naphthalene). Adsorbed organosilicon compounds can be removed by elution with methyl isobutyl ketone (ref. 143). Synachrom (manufactured by Lachema, Brno, Czechoslovakia) is a copolymer with properties very similar to Porapak Q. The sorption process is physical sorption; thus, sorption on this copolymer is inversely proportional to solubility of sorbates in water.

Porapak N is not a very good copolymer for clean-up purposes. Comparing it to Porapak Q it is found that it possesses no ability to trap silicon compounds (ref. 143).

TABLE 4.5

Comparison of various types of amberlite (ref. 144)

Compound	Concentration (ppb)	Recovery(%) for XAD resin type										Ref.
		7	2/4	8	2/7	2	2/8	2/4/7/8	4	4/8	12	
Acenaphthene	50	50	72	68	20	81	99	85	84	81	81	145
2-benzothiazole		40	80	53	75	74	73	77	82	82		
Bis(2-chloroisopropyl)ether		-	74	77	76	76	77	71	80	82		
p-Cresol		33	58	47	50	44	49	60	69	68		
Dibenzofuran		73	70	95	83	93	86	85	82	84		
n-Hexadecane		3	-	-	8	3	18	14	-	11		
1-Methylnaphthalene		64	62	80	75	76	82	80	77	79		
2-Methylnaphthalene		63	61	77	72	75	80	81	77	77		
o-Nitrotoluene		53	75	77	79	82	81	81	83	83		
Naphthalene		64	66	78	77	79	81	82	80	80		
Phenol		19	30	29	32	14	33	41	38	46		
-Terpineol		36	76	62	77	81	75	75	80	80		
sym-Tetrachloroethane		35	58	59	66	61	68	68	72	72		
Dehydroabietic acid	100	31	-	85	-	94	-	-	86	-		145
Di-2-ethylhexyl phthalate		22	-	13	-	33	-	-	11		-	
2-Ethylhexanol		74	-	79	-	85	-	-	91		-	
Isophorone		46	-	47	-	76	-	-	86		-	
Palmitic acid		12	-	16	-	67	-	-	79		-	
Pentachlorophenol		83	-	77	-	84	-	-	84		-	
-BHC	1	-	-	28	-	53	-	-	90	-	9	146
Lindane		-	-	17	-	45	-	-	107	-	10	
-BHC		-	-	-	-	71	-	-	81	-	32	
Aldrin		-	-	63	-	51	-	-	50	-	25	
Dieldrin		-	-	82	-	61	-	-	107	-	25	
Cumene	10,000	67	-	-	-	67	-	-	-	-	-	147
Ethylbenzene		59	-	-	-	60	-	-	-	-	-	
Naphthalene		93	-	-	-	90	-	-	-	-	-	
n-Hexane		83	-	-	-	82	-	-	-	-	-	
Phenol		45	-	-	-	27	-	-	-	-	-	
Octanoic acid		81	-	-	-	58	-	-	-	-	-	
o-Cresol		67	-	-	-	59	-	-	-	-	-	
Chlorophenol		85	-	-	-	70	-	-	-	-	-	

(by permission of publisher.)

Another group of polymers which may be used is the Chromosorb Century Series. Chromosorb 105 is a polyaromatic type polymer having an average polarity, and average pore diameter of 0.04-0.06 μm and specific surface area of 600-700 m^2/g. This polymer is not recommended for clean-up properties. Chromosorb 106 is very similar to the Chromosorb 105.

Table 4.5 illustrates a comparison of the various types and/or mixtures of the Amberlite XAD resins. The data show that a 1:1 mixture of XAD-4 and XAD-8 resins is the most efficient and that resin XAD-4 is very efficient for chlorinated insecticides. XAD-7 is a better sorbent than XAD-2 for polar compounds, because XAD-7 has a higher polarity. XAD-8 is most suitable (as compared to XAD-2) for natural water studies because it has less tendency to become clogged by natural polyelectrolytes (ref. 144).

A comparison of the recoveries with XAD-2 and activated charcoal (Filtrasorb 300) was made by Chriswell et al. (ref. 148). These data are shown in Table 4.6. The data show that the porous polymer, XAD-2, is more suitable than charcoal for recoveries of the various sorbates except for acidic type compounds (neither sorbent is acceptable) or alkanes. A similar study between Amberlite XAD-4 and Spherocarb (carbon molecular sieve) furnished essentially the same information (ref. 131). The adsorption capacity of carbon is lower than Amberlite XAD resins, Synachrom and open-pore polyurethanes (ref. 10).

TABLE 4.6

Comparison of recoveries with Amberlite XAD-2 and activated carbon (ref. 148) (by permission of publisher)

Compound	Number of investigated compounds	Recovery(%) Resin	Carbon
Alkanes	5	5	15
Esters	4	61	49
Alcohols	8	73	47
Phthalate esters	3	82	24
Phenols	10	45	7
Chlorinated alkanes and alkenes	5	43	55
Chlorinated aromatic compounds	13	70	11
Aromatic compounds	7	68	6
Aldehydes and ketones	3	74	4
Amines	13	54	54
Carboxylic acids	11	1	2
Pesticides	4	34	16
Miscellaneous	14	33	11

Marcelin (ref. 149) used Tenax-GC as a clean-up packing when analyzing mixtures of permanent gases and heavy organic compounds. A short pre-column of Tenax-GC adsorbed the organic material and passed the permanent gases and water into a column of Chromosorb 101 folled by a column of Porapak QS for separation. Thermal desorption of the organic material and separation on a 10% XE-60/Chromosorb WHP column followed. A clean-up and concentration system for chlorinated insecticide residues has been described by Solomon (ref. 150). Hexane solutions of the chlorinated pesticides were chromatographed on a Florisil column and eluted with hexane-diethyl ether solvent. Recoveries of 87-100% were reported. Baird et al. (ref. 151) carried out a study on the effluents from four pilot tertiary waste-water treatment systems for trace organic compounds. The effects of ozone and free and combined chlorine were also studied.

They used various combinations of alum flocculation, filtration, and carbon adsorption for removal of volatile halogenated organics, polynuclear aromatic hydrocarbons, chlorinated pesticides, and polychlorinated biphenyls. Gas chromatography was the method of analysis. Carbon adsorption reduced the trihalomethane levels almost 90%. Chang and Fritz (ref. 152) used macroporous XAD-2 resin to concentrate and then determine various pollutants in water. Average recovery was 83% at the 3-200 ppb (10^9) level, with an overall reproducibility of 5-8%.

4.5 DERIVATIZATION

Derivatization is the process of chemically modifying a molecule in order to make it more compatible with an analytical procedure. This is not a process unique to chromatographic separation. Derivatizations are performed for a variety of reasons:

(1) To increase volatility of a component or to increase its thermal stability for gas chromatography, i.e., improve the chromatographic behavior of the component. An example would be the silylation or acetylation reactions of hydroxyl or carboxylic acid groups. In liquid chromatography this may be used to change the polarity of the parent molecule, e.g., conversion of alcohols to ethers or carboxylic acids to esters. The sensitivity in liquid chromatography may be increased by forming a nitro-substituted ester of the carboxylic acid. The molar adsorptivity of the nitro-substituted ester is greater than the simple ester. The making of a fluorometric derivative prior to the separation allows the use of a fluorometric detector and increases the sensitivity a hundred-fold (ref. 153).

(2) To facilitate the separation of components in a mixture: The parent molecule may be changed to another chemical form which has a retention time different from the parent molecule. This technique may be used in gas chromatography as well as in liquid chromatography; however, its use in liquid chromatography is limited because of the variety of solvent systems available. In this application the derivative need not give a more sensitive signal than the parent molecule. Very polar compounds, e.g., free acids or amines, are difficult to separate by gas chromatography as are some thermally labile compounds. The chemical modification of the functional group improves the chromatographic behavior of the molecules (its ability to be determined by gas chromatography).

(3) To increase the detectability by imparting a measurable characteristic to the parent molecule: Pollutants are usually present in the ppm or ppb range and one may increase the sensitivity of the method by forming a halogen or acyl derivative (tagging with a perfluoro alkyl or acyl group). In this manner the detection can be carried out by use of an electron-capture detector. A minimal resultant increase in sensitivity by a factor of a thousand is often obtained. Pesticide residues in food or soils may easily be detected in this manner. One may extend the utility of liquid chromatography detectors by tagging molecules of interest with strong UV chromophores

or with a fluorophore (for fluorescent detectors). There are a number of reagents available for making derivatives of molecules to enhance their detectability. These reagents may be grouped according to the type of parent molecule and reagent type. Parent molecules having a replaceable hydrogen easily react, e.g., amines, alcohols and phenols. Halogenated trimethylsilyl (TMS) derivatives are very good. Also trifluoroacetic anhydride (TFAA), pentafluoropropionic anhydride (PFPA), and heptafluorobutyric anhydride (HFBA) are commonly used. These derivatives not only increase the detectability but also aid in volatility and stability. Alcohols and amines form derivatives with N-trifluoroacetylimidazole and N-heptafluorobutyrimidazole. A very good reagent for primary and secondary amines is pentafluorobenzoylchloride.

If derivatization is to be used, one must decide if the chemical transformation will take place before injection onto the column (pre-column derivatization), during the separation process in the column (on-column derivatization) or at the completion of the separation (post-column derivatization).

(a) Pre-column derivatization is used when the derivative formation reaction time is long or may require drastic conditions. The chemical reaction may take place overnight followed by gas or liquid chromatographic separation of the product.

(b) On-column derivatization may make use of the heated injection port and/or column (in gas chromatography) to effect the chemical reaction. Decision making on the system can not wait until the next day because derivatives made this way must have fast reaction time and be a clean reaction. This technique is preferred over pre-column derivatization because of the speed and simplicity. These types of derivatives (gas chromatography) mean modification to system equipment. Unless these procedures are to be used on a routine basis, one should consider doing pre-column derivatization.

(c) Post-column derivatization is primarily used in liquid chromatography rather than in gas chromatography so that the detectability of the components is enhanced. This involves the separation of parent compounds and the detection of the derivative; whereas, with a and b, the separation and detection are based upon the derivative. This mode of derivatization introduces the possibility of mixing of the separated components of the sample after they exit from the column and before they enter the detector.

In all three techniques, matrix effects can interfere. Thus it is essential to be certain of the qualitative make-up of the sample. One parameter which can cause an interference is pH, because the change in hydrogen ion concentration may inhibit the reaction of interest or accelerate an unwanted reaction.

A reagent to be useful for derivatization should meet certain criteria:

(1) It should form no more than one derivative with each parent compound.

(2) Reaction completion should require minimum time under mild conditions.

(3) Any by-products or excess reagent should not interfere with the separation of the compounds of interest. In other words, reaction by-products should be either non-UV adsorbing or sufficiently different in structure from the derivative of interest that separation is no problem.

(4) Differences among the parent compounds should be preserved to allow their separation.

Substituted aromatic compounds are the most commonly used UV chromophores for liquid chromatography derivatives. They represent the best general compromise among size, polarity, and molar adsorptivity.

The most commonly used derivative reagent for gas chromatography has been the silyl ether. The reaction between sample component and silylation reagent is fast and leaves a pure derivative in fairly high yield. A silyl derivative may be made if the sample component has a functional group with silyl acceptor ability. Most compounds with oxygenated or nitrogen-containing functional groups will form this type derivative. One can rank these functional groups in order of their ability to react with a silyl reagent. Alcohols are more reactive than phenols which are more reactive than carboxylic acids, which are better than amines and amides. Within a functional grouping the order of reactivity is primary > secondary > tertiary because steric factors play an important role in the formation of derivatives.

There is a wide variety of derivatives available for gas chromatography and liquid chromatography. Many of these are for compounds which are of little interest to the environmental chemist (e.g., carbohydrates and amino acids) and so will not be discussed here. For a general discussion of derivatization see the book by Blau and King (ref. 154), the book by Lawrence and Frei (ref. 155), and the special issue of the Journal of Chromatographic Science edited by Lawrence (ref. 156).

At the present time, the making of derivatives is not often performed for pollutants. The pollutants which are of the most concern can easily be determined by gas chromotography, gas chromotography/mass spectrometry, or liquid chromotography. However, this does not rule out the possibility of determining by derivitization pollutants which may be added to the various priority pollutant lists throughout the world or the derivatization of pollutants to increase detectability. We have tabulated derivatives which may be used at this time (see Table 4.7), with the projection that higher molecular weight or less volatile pollutants will become of interest. If this comes to pass, derivatives which are more volatile than the parent compound will be needed. Another justification for derivatives will be for the detection of pollutants if the maximum allowable limits are lowered. If this occurs, the analytical chemist will need more sensitive means of detection. Incorporating functional groups into a parent pollutant molecule may make it more easily detected (e.g., halogenated derivatives so that smaller amounts may be detected by electron-capture detectors in lieu of flame ionization detectors).

The formation of derivatives can impose some penalties which should be considered. First of all the time for an analysis will increase thus increasing the cost per sample analyzed. Secondly, additional reagents will be necessary to form the derivatives - again increasing the cost per sample analyzed. Thirdly, additional equipment will become necessary with its additional costs.

Although most pollutants are easily determined at the present time, the use of a derivative and its subsequent determination is a straightforward technique for confirming the presence of a particular pollutant. This is especially true for those laboratories which do not have mass spectrometry capabilities.

Sulfur-containing acids, in atmospheric aerosols, have been methylated with diazomethane and the resulting esters separated on a diethylene glycol succinate column and determined with a sulfur-sensitive detector (ref. 214). A similar technique converted carbamate pesticides to sulfonic esters and separated them on a LSX-3-1295 or W-98 on CHromosorb W column 220°C (ref. 122). Nitrogen was the carrier gas and detection was with either electron-capture or sulfur-mode flame photometric detectors. Solomon and Kallas (ref. 123) separated and quantified 2,4,6-trichlorophenyl ether derivatives at sub-ppb levels using a 0.1% QF-1 + 0.1% OV-17 gas-liquid chromotographic column and electron-capture detection. Chlorinated phenols may be derivatized with diazomethane or diazoethane or a silanizing agent and then determined at ng/ml (ppb) levels using electron-capture detection (ref. 215). If the phenol does not contain a group with electron capture response, heptafluorobutyrate may be used as the derivatizing agent. Recoveries of the fluorinated phenol derivatives at the 20-200 ppb level were greater than 75%. Giam et al. (ref. 216) confirmed the identity of phthalate esters by formation of the N-(2-chloroethyl)phthalimide derivatives followed by separation on a 3% SE-30 column and detection with a ^{63}Ni electron-capture detector. Cochrane (ref. 217) has reviewed the use of chemical derivatization of pesticides prior to gas chromatographic or high-performance liquid chromatographic analysis. In gas chromatography the pesticides are derivatized to more stable and volatile products; while in high-performance liquid chromatography, derivatives are made to increase detectability. In high-performance liquid chromatography, one introduces either a chromophoric or fluorophoric group into the molecules. This lowers the detection limits below the ppm level.

Cyanide ion may be detected by gas liquid chromatography by converting it to cyanogen chloride using chloramine T (ref. 218). A packed column made from 7% Hallcomid M-18 on Anakrom ABS is used for the separation and an electron-capture detector used for quantitation. Detection limits are 25 ng/ml (25 ppb). Metals in environmental samples may be determined by gas liquid chromatography if first converted to metal chelate complexes which are stable. Fluorine substituted ∞-diketones are extremely sensitive to electron-capture detection and thus the chelating agent of choice. The hexafluoromonothioacetyl acetonate complexes of Cu(II), Zn(II), Ni(II), Cd(II), and Pb(II) have been prepared. Detection limits range from the nanogram to the picogram levels (ref. 219).

TABLE 4.7

Derivatization used in gas and liquid chromatography

Pollutant type	Derivative and/or derivative reagent	Comments	Ref.
Acids(carboxylic)	o-p-Nitrobenzyl-N,N'-diisopropylisourea (PNBDI)	Liquid chromatography pre-column derivative. Long-chain acids separated from short-chain acids by adsorption.	179
	p-Bromophenacyl bromide (PBPB) with 18-Crown-6(1,4,7,10,13,16-hexaocxcyclo-octadecane).	Liquid chromatography pre-column derivative. Resulting solution may be chromotographed directly.	180
Acids (Fatty)	bis-Trimethyl-silyltrifluoroacetamide(BSTFA) + trimethylchlorosilane (TMCS).	Gas chromatography derivative-pre-column.	157
	4-Bromomethyl-7-methoxy coumarin (BMC)	Liquid chromatography pre-column derivative.	181
Acids (phenolic)	BSTFA + TMCS	Gas chromatography derivative pre-column.	157
Alcohols	Trimethyl-silyl (TMS)ether	Gas chromatography derivative-pre-column.	158-162
	3,5-Dinitrobenzoyl chloride(DNBC)	Liquid chromatography, no further workup needed,	155,156 163,164

Pollutant type	Derivative and/or derivative reagent	Comments	Ref.
		pre-column derivative.	
	Pyruvoyl chloride (2,6-dinitrophenyl) hydrazone	Liquid chromatography, no further workup needed, column derivative.	165
	p-Iodobenzenesulfonylchloride	Liquid chromatography, requires further clean-up, pre-column derivative.	165
	Benzoyl chloride	Liquid chromatography, pre-column derivative, requires further clean-up.	166
	p-Nitrobenzoyl chloride	Liquid chromatography, pre-column derivative, requires further clean-up.	167
Aldehydes and ketones	Oxime	Gas chromatography pre-column derivative.	168
	Methoxime	Gas chromatography pre-column derivative.	169
	Phenylhydrazone	Gas chromatography pre-column derivative.	170
	Dinitrophenyl-hydrazine	Gas chromatography pre-column derivative.	171-173
	p-Nitrobenzyl-oxyamine HCL(PNBA)	Liquid chromatography pre-column derivative.	163,174
	Dansyl hydrazine	Liquid chromatography pre-	181
	(1-dimethylamino naphthalene-5-sylfonyl hydrazine)	column derivative.	

Pollutant type	Derivative and/or derivative reagent	Comments	Ref.
Amines	BSTFA	Gas chromatography pre-column derivative, 90°C for 15 min, ethyl acetate solvent.	175
	N-DNP	Gas chromatography pre-column derivative.	176-178
	N-Dimethylamino-methylene(DMAM)	Gas chromatography pre-column derivative.	182
	N-TMS	Gas chromatography pre-column derivative.	183
	N-Trifluoro-acetamide(TFA)	Gas chromatography pre-column derivative	184-188
	N-Heptafluoro-butyric acid(HFB)	Gas chromatography pre-column derivative.	160,189-191
	N-Pentafluorophenyl-ester(N-PF Phenyl)	Gas chromatography pre-column derivative	189,192-194
	N-Pentafluoro-benzylidine(N-PF) Benzylidine).	Gas chromatography pre-column derivative	189,192-194
	N-Pentafluoro-benzamide(N-PF Benzamide)	Gas chromatography pre-column derivative.	189,192-194
	N-Pentafluoropro-pionyl(PFP)	Gas chromatography pre-column derivative.	195-198
	PFB-Carbamate	Gas chromatography pre-column derivative.	199-201
	Hoffman degradation	Gas chromatography pre-	202

Pollutant type	Derivative and/or derivative reagent	Comments	Ref.
		column derivative.	
Amines (phenolic)	CS_2 with $-NH_2$ group. BSTFA with -OH group.	Gas chromatography pre-column derivative.	175
	DNBC	Liquid chromatography derivative, clean-up not necessary, pre-column derivative.	155,163
	2,6-(Nitrophenyl) hydrazone	Liquid chromatography pre-column derivative, no clean-up necessary.	165
	p-Methoxybenzoyl chloride	Liquid chromatography pre-column derivative, clean-up necessary before sample injection.	203
Amines (primary + secondary)	N-Succinimidyl-p-nitrophenyl acetate (SNPA)	Liquid chromatography pre-column derivative.	163
	2,4-Dinitro-1-fluorobenzene	Liquid chromatography pre-column derivative, clean-up necessary before sample injection.	155
	5-N,N-dimethylamino-naphthalene-1-sulfonyl chloride(DNS chloride)	Liquid chromatography pre-column derivative, clean-up necessary.	155
Amines (primary)	DNS chloride	Reaction carried out in alkaline buffer. Liquid chromatography pre-column derivative.	181

Pollutant type	Derivative and/or derivative reagent	Comments	Ref.
Amines (primary and secondary)	7-Chloro-4-nitro-benzo-2-oxa-1,3-di-azole(NBD chloride)	Liquid chromatography, pre-column derivative. Formed under mildly alkaline conditions.	181
Herbicides	Trimethylanilinium hydroxide (TMAH)	Gas chromatography pre-column derivative. Also on-column with high temperature injection port.	204
Isocyanate monomers	p-Nitrobenzyl-N-n-propylamine hydro-chloride(PNBPA)	Liquid chromatography pre-column derivative. Detection limits of 0.7ppb(v/v)toluene isocyanate in 20-liter air sample.	205
Pesticides	BSFTA	Gas chromatography pre-column derivative.	206,207
N-methylcar-bamate in-secticides.	o-Phthaldehyde(OPA)	Liquid chromatography post-column derivative.	213
Phenols	TMS ether	Gas chromatography pre-column derivative.	161, 208-212
	DNBC	Liquid chromatography pre-column derivative, no deriva-tive clean-up necessary before injection.	155,163
	2,6-Nitrophenyl hydrazone	Liquid chromatography pre-column derivative, no clean-up necessary.	165
	p-Iodobenzenesul-fonylchloride	Liquid chromatography pre-column derivative, clean-up necessary before injection.	165

Pollutant type	Derivative and/or derivative reagent	Comments	Ref.
	DNS chloride	Liquid chromatography pre-column derivative, clean-up necessary before injection.	155
Thiols	NBD chloride	Liquid chromatography pre-column derivative.	181

REFERENCES

1 R. D. Cadle, P. L. Magill, A. A. Nichol, H. C. Ehrmantraut and G. W. Newell, Sampling Procedures, in P. L. Magill, F. R. Holden and C. Ackley (Eds.), Air Pollution Handbook, McGraw-Hill, New York, 1956, Ch. 10.
2 J. W. Miles, E. W. Dale and F. C. Churchill, Arch. Environ. Contam. Toxicol., 5 (1976) 29; Anal. Abstr., 33 (1977) 1H59.
3 P. M. Giever, Particulate Matter Sampling and Sizing, in A. C. Stern (Ed.), Air Pollution, Academic Press, New York, 1976, Ch. 1.
4 L. D. Kissinger and J. S. Fritz, J. Amer. Water Works Ass., 68 (1976) 435.
5 A. A. Rosen and F. M. Middleton, Anal. Chem. 31 (1959) 1729.
6 F. M. Middleton and J. J. Litchtenberg, Ind. Eng. Chem., 52 (1960) 99A.
7 R. A. Baker, J. Amer. Water Works Ass., 58 (1966) 751.
8 H. Braus, F. M. Middleton and G. Walton, Anal. Chem. 23 (1951) 1160.
9 G. F. Lee, G. W. Kumke and S. L. Becker, Int. J. Air Water Pollut., 9 (1965) 69.
10 M. Dressler, J. Chromatogr., 165 (1979) 167.
11 W. Nernst, Z. Phys. Chem., 8 (1891) 110.
12 Methods for Organic Compounds in Municipal and Industrial Wastewaters, U. S. EPA Environmental Monitoring and Support Laboratory, Cincinnati, OH March 1979.
13 R. L. Grob, Progress in Priority Pollutant Analysis, Special Issue, J. Environ. Sci. Health, Part A, Environ. Sci. Eng., A15 (1980) 379-543.
14 R. A. Hites, Advan. Chromatogr. 15 (1977).
15 T. W. May and D. L. Stalling, Anal. Chem. 51 (1979) 169.
16 R. Dell'Acqua, J. A. Egan and B. Bush, Environ. Sci. Technol., 9 (1975) 38.
17 D. B. Harper, R. V. Smith and D. M. Gatto, Environ. Polln., 12 (1977) 223.
18 E. B. Overton, J. Bracken and J. L. Laseter, J. Chromatogr. Sci., 15 (1977) 169.
19 N. T. M. Kleverlaan, Chem. Weekbl., 71 (1975) 13; Anal. Abstr., 30 (1976) 5H34.
20 Z. I. Chalaya, L. S. Mikhailova, A. I. Mashkevich, O. A. Egorerichenko and O. M. Polichchuk, Zh. Anal. Abstr., 30 (1976) 5H43.
21 I. S. Kofman and V. I. Kafanov, Gig. Sanit., No. 10, (1979) 41; C.A., 92 (1980) 35763c.
22 G. L. LeBel and D. T. Williams, J. Ass. Offic. Anal. Chem., 62 (1979) 1353.
23 P. F. Blanchet, J. Chromatogr., 179 (1979) 123.
24 D. L. Stalling and J. N. Huckins, U.S. Environ. Prot. Agency Off. Res. Dev. Rep. EPA-600/3-76-076 (1976); Anal. Abstr., 33 (1977) 1D66.
25 J. J. Franken and B. J. M. Lyten, J. Ass. Offic. Anal. Chem., 59 (1976) 1279.
26 J. P. Cane, J. Guintrand, C. Aubert and A. Viala, Arzeneim.-Forsch., 27 (I)(2) (1977) 338; Anal. Abstr., 33 (1977) 2D39.
27 M. Galoux, J. C. VanDamme, A. Bernes and J. Potvin, J. Chromatogr., 177 (1979) 245.
28 W. Bates, K. Wilcopolski and B. Lehmann, Lebensmittelchem, Gerichtl. Chem., 33, No. 4 (1979) 73; C.A. 91 (1979) 191465s.
29 J. Singh, W. P. Cochrane and J. Scott, J. Bull. Environ. Contam. Toxicol., 23 (1979) 470.

30 D. A. J. Murray, J. Chromatogr., 177 (1979) 135.
31 K. Grob, K. Grob, Jr. and G. Grob, J. Chromatogr., 106 (1975) 299.
32 J. W. Rhoades and C. P. Nulton, J. Environ. Sci. Health, Part A, Environ. Sci. Eng., A15 (1980) 467.
33 C. B. Koons, Proc. 1977 Oil Spill Conference, American Petroleum Institute, Washington, DC, 1978, 589-591.
34 C. D. McAuliffe, Science 158 (1969) 478.
35 C. D. McAuliffe, Chem. Technol. 1 (1971) 46.
36 P. D. Wagner, P. F. Naumann and R. B. Laravuso, J. Appl. Physiol. 36 (1974) 600.
37 C. D. McAuliffe, U.S. Patent 3,345,137, Oct. 3, 1967.
38 J. W. Swinnerton and V. J. Linnenbom, J. Gas Chromatogr., 5 (1967) 570.
39 K. Grob, J. Chromatogr., 84 (1973) 255.
40 J. Novak, J. Zululeckey, V. Kubelka and J. Mostecky, J. Chromatogr., 76 (1973) 45.
41 K. Grob and F. Zurcher, J. Chromatogr., 117 (1976) 285.
42 J. W. Swinnerton and V. J. Linnenbom, Science, 156 (1976) 1119.
43 C. D. McAuliffe, J. Phys. Chem., 10 (1966) 1267.
44 L. C. Price, Bull. Amer. Ass. Pet. Geol., 60 (1976) 213.
45 M. Kojima, Nippon Shokuhin Kogyo Gakkaishi, 24, No. 2 (1977) 90; C.A. 91 (1979) 156172x.
46 L. E. Reshetnikova, L. I. Slyusareva, T. N. Shivalova, I. I. Prusakova and N. T. Karabanov, Fiz. Khim. Metody Anal. 3 (1978) 98.
47 I. Y. Nikitin and O. V. Klipikova, Deposited Doc., VINITI, (1978) 1054; C.A., 91 (1979) 162471k.
48 G. S. Fedchenko and M. S. Vigdergauz, Khim. Tekhnol. Topl. Masel., No. 7 (1979) 51; C.A., (1979) 163477u.
49 P. Zimmerman and R. Rasmussen, Environ. Sci. Technol., 9 (1975) 1077.
50 U. Cernot, Clin. Chim. Acta, 94 (1979) 48; C.A. 91 (1979) 119722f.
51 P. J. Noomen, Chem. Mikrobiol. Technol. Lebensm., 6 No. 2 (1979) 48; C.A., 91 (1979) 122254k.
52 B. Kolb, P. Pospisil, T. Borath and M. Auer, J. High Resolut. Chromatogr. Chromatogr. Commun., 2 (1979) 283.
53 R. J. Rath, D. Schmidt and J. Wimmer, Chromatographia, 12 (1979) 567.
54 A. B. Danilovtseva, Y. D. Kadaner and K. A. Kalunyants, Izv. Vyssh. Uchebn. Zaved. Pishch. Tekhnol., (1979) 122; C.A., 91 (1979) 54535a.
55 N. A. Tarasova and S. E. Kataeva, Gig. Sanit, No. 3 (1979) 48; C.A., 91 (1979) 54648q.
56 J. Drozd and J. Novak, J. Chromatogr., 165 (1979) 141.
57 E. A. Dietz, Jr. and K. F. Singley, Anal. Chem., 51 (1979) 1809.
58 K. L. E. Kaiser and B. G. Oliver, Anal. Chem.,48 (1976) 2207.
59 R. M. Maiorino, I. G. Sipes, A. J. Gandolfi and B. R. Brown, Jr., J. Chromatogr., 164 (1979) 63.
60 V. V. Tsibul'skii, I. A. Tsibul'skaya, and N. N. Yaglitskaya, Zh. Anal. Khim., 34 (1979) 1365.
61 G. G. Rusakova and I. A. Kirpa, Khim. Prom-st., Ser.: Methody Anal. Kontrolya Kach. Prod. Khim. Prom. sti., No. 7 (1979) 10; C.A., 92 (1980) 33426w.
62 B. V. Ioffe, A. G. Vitenberg and I. A. Tsilbul'skaya, J. Chromatogr., 186 (1979) 851.
63 J. W. Elkins, Anal. Chem., 52 (1980) 263.
64 J. Drozd and J. Novak, J. Chromatogr., 152 (1978) 55.
65 R. C. Huber, A. O. Niedermayer and A. L. Weiss, J. Pharm. Sci., 67 (1978) 1239.
66 E. L. Thomas, G. A. Reineccius, G. J. DeWaard and M. S. Swinkard, J. Dairy Sci., 59 (1976) 1865; Anal. Abstr., 33 (1977) iF20.
67 D. T. Williams, J. Ass. Offic. Anal. Chem., 59 (1976) 30.
68 A. Jabbagy and J. Hallo, Nahrung, 20 (1976) 287; Anal. Abstr., 31 (1976) 4F32.
69 G. B. M. Gawell, Analyst (Londen), 104 (1979) 106.
70 K. W. Grubaugh and G. E. Stobly, Anal. Chem., 50 (1978) 377.
71 T. Tusuneya, N. Ikeda, M. Shiga and N. Ichikawa, Int. Congr. Essent. Oils (Pap) 7 (1977 Pub 1979), 454-457.
72 J. S. Smith, D. T. Burkett, and J. M. Hanrahan, ASTM Spec. Tech. Publ., 1978, STP 686, Measurement Organic Pollution Water, Waste Water, 1979, pp. 251-255.

73 G. M. Loper and J. L. Webster, J. Chromatogr. Sci., 9 (1971) 466.
74 R. Teranishi, T. R. Mon, A. B. Robinson, P. Cary and L. Pauling, Anal. Chem., 44 (1972) 18.
75 A. B. Robinson, D. Partridge, M. Turner, R. Teranishi and L. Pauling, J. Chromatogr., 85 (1973) 19.
76 R. E. Hurst, Analyst (Londen), 99 (1974) 302.
77 R. E. Hurst, Anal. Chem., 47 (1975) 1221.
78 R. Bassette, S. Ozeris and C. H. Whitnah, Anal. Chem., 34 (1962) 1540.
79 V. Palo, Chromatographia, 4 (1971) 55.
80 M. Gottauf, Z. Anal. Chem., 218 (1966) 175.
81 H. G. Maier, J. Chromatogr. 50 (1970) 329.
82 P. L. Davis, J. Chromatogr. Sci., 8 (1970) 423.
83 P. J. Gilliver and H. E. Nursten, Chem. Ind. (Londen) (1972) 541.
84 H. Binder, Z. Anal. Chem., 244 (1969) 353.
85 H. Hachenberg, TIPS, 41 GC, Bodenseewerk Perkin-Elmer Uberlingen, G.F.R., April 1970.
86 S. Ozeris and R. Bassette, Anal. Chem., 35 (1963) 1091.
87 R. E. Kepner, H. Maarse and J. Strating, Anal. Chem., 36 (1964) 77.
88 D. Jentzch, H. Kruger and G. Lebrecht, Angewandte Gaschromatographia, 19, Part 9 Bodenseewerk Perkin-Elmer, Uberlingen, G.F.R.,1967.
89 H. Hachenberg, Industrial Gas Chromatographic Trace Analysis, Heyden, Philadelphia, London, 1978, pp. 108-123.
90 H. Hachenberg and A. P. Schmidt, Gas Chromatographic Headspace Analysis, Heyden, Philadelphia, London, 1977.
91 G. R. Umbreit and R. L. Grob, Environ. Sci. Health, Part A, Environ. Sci. Eng., A15 (1980) 429.
92 J. Drozd and J. Novak, J. Chromatogr., 136 (1977) 37.
93 J. Drozd, J. Novak and J. A. Rijks, J. Chromatogr., 158 (1978) 471.
94 W. J. Khazal, J. Vejrostra and J. Novak, J. Chromatogr., 157 (1978) 125.
95 T. A. Bellar and J. J. Lichtenberg, J. Amer. Water Works Ass., 66 (1974) 739.
96 T. A. Bellar and J. J. Lichtenberg, Semi-automated Headspace Analysis of Drinking Waters and Industrial Waters for Purgable Volatile Organic Compounds, U.S. EPA, Environmental Monitoring and Support Laboratory, Cincinnati, OH, June 1978.
97 The Analysis of Trihalomethanes in Finished Waters by the Purge and Trap Method, Method 501.1, U.S. EPA, Environmental Monitoring and Support Laboratory, Cincinnati, Oh, May 15, 1979.
98 D. Beggs, Amer. Lab., July (1978) 81-87.
99 W. D. Snyder, Hewlett-Packard Technical Paper GC-71, Hewlett-Packard, Avondale, PA 1971.
100 J. Watkins and J. Poole, Hewlett-Packard Technical Paper 78, Hewlett-Packard, Avondale, PA 1978.
101 R. R. Freeman, T. Rooney, T. Przybski and L. Altmeyer, Hewlett-Packard Technical Paper 83, Hewlett-Packard, Avondale, PA 1980.
102 B. Kolb, J. Chromatogr., 122 (1976) 553.
103 B. Kolb, J. Chromatogr., 112 (1975) 287.
104 B. Kolb, E. Wiedeking and B. Kempken, Angewandte Gas-Chromatographie, Bodenseewerk Perkin-Elmer, Uberlingen, G.F.R.,
105 Perkin-Elmer Technical Bulletin, Model F42 Head Space Analyzer, MP8/75 10, Perkin-Elmer, Norwalk, CT, August 1975.
106 M. Mikaus and A. Vanko, Petrochemia, 19, No. 3 (1979) 49; C.A. 92 (1980) 42574v.
107 W. F. Cowen, W. J. Cooper and J. W. Highfill, Anal. Chem. 47 (1975) 2483.
108 C. D. McAuliffe, Symposium on Environmental Analysis, Fall 1978 ACS Meeting, Miami Beach, FL, 1978.
109 Development and Application of Test Procedures for Specific Organic Toxic Substances in Wastewaters, Category 11-Purgable and Category 12-Acrolein, Acrylonitrile and Dichlorodifluoromethane, Report for EPA Contract 68-03-2635, Cincinnati, OH, 1979.
110 Purgeable Halocarbons, Method 601, U.S. EPA, Environmental Monitoring and Support Laboratory, Cincinnati, OH, April 1979.
111 Purgeable Aromatics, Method 602, U.S. EPA, Environmental Monitoring and Support Laboratory, Cincinnati, OH, April 1979.

112 Acrolein and Acrylonitrile, Method 603, U.S. EPA, Environmental Monitoring and Support Laboratory, Cincinnati, OH, April 1979.
113 J. W. Swinnerton, V. J. Linnenbom and C. H. Cheek, Anal. Chem., 34 (1962) 1509.
114 J. W. Swinnerton, V. J. Linnenbom and C. H. Cheek, Anal. Chem., 34 (1962) 483.
115 J. W. Swinnerton, V. J. Linnenbom and C. H. Cheek, Environ. Sci. Technol., 3 (1969) 836.
116 S. P. Wasik and W. Tsang, Anal. Chem., 42 (1970) 1649.
117 J. O. Grote and R. G. Westendorf, Amer. Lab., Dec. (1979) 61-66.
118 R. A. Hites, Advan. Chromatogr., 15 (1977) 69-112.
119 T. A. Bellar, J. L. Lichtenberg and J. W. Eichelberger, Environ. Sci. Technol., 10 (1976) 926.
120 R. Mindrup, Jr., NBS Spec. Publ. (U.S.), 519 (1979).
121 B. J. Dowty, S. R. Antoine, and J. L. Laseter, ASTM Spec. Publ., 1978, STP 686. Measurement Organic Pollution Water, Wastewater, 1979, pp. 24-35.
122 H. A. Moye, J. Agr. Food Chem., 23 (1975) 415; Anal. Abstr. 30 (1976) 1G20.
123 R. A. Solomon and G. J. Kallas, Anal. Chem., 47 (1975) 955.
124 S. Glasstone, Textbook of Physical Chemistry, D. van Nostrand, New York, 2nd ed., 1946, pp.700, 729, 967, 1254.
125 A. A. Rosen and E. M. Middleton, Anal. Chem., 31 (1959) 1729.
126 J. W. Eichelberger and J. L. Lichtenberg, J. Amer. Water Works Ass., 63 (1971) 25.
127 C. D. Chriswell, R. C. Chang and J. S. Fritz, Anal. Chem., 47 (1975) 1325.
128 L. R. Snyder, Principles of Adsorption Chromatography, Marcel Dekker, New York, 1968, p. 192.
129 C. D. Chriswell, L. D. Kissinger and J. S. Fritz, Anal. Chem., 48 (1976) 1123.
130 G. A. Junk, J. J. Richard, M. D. Gresser, D. Witiak, J. L. Witiak, M. D. Arguello, R. Vick, H. J. Svec, J. S. Fritz and G. V. Calder, J. Chromatogr., 99, (1974) 745.
131 A. Tateda and J. S. Fritz, J. Chromatogr., 152 (1978) 329.
132 J. P. Riley and D. Taylor, Anal. Chim. Acta, 46 (1969) 307.
134 K. Salodynskii, L. Panina and N. Klinskaya, Chromatographia, 7 (1974) 339.
135 V. Leoni, G. Pucceti and A. Grella, J. Chromatogr., 106 (1975) 119.
136 B. Versino, H. Knoppel, M. DeGroot, A. Peil, J. Poelman, H. Schauenburg, H. Vissers and F. Geis, J. Chromatogr., 122 (1976) 373.
137 J. D. Navrotil, R. F. Sievers and H. F. Walton, Anal. Chem., 49 (1977) 2260.
138 H. D. Gesser, A. Chow, F. C. Davis, J. F. Uthe and J. Reinke, Anal. Lett., 4 (1971) 883.
139 J. Saxena, J. Kozuchowski and D. K. Basu, Environ. Sci. Technol., 11 (1977) 682.
140 A. K. Burnham, G. V. Calder, J. S. Fritz, G. A. Junk, H. J. Svec and R. Willis, Anal. Chem., 44 (1972) 139.
141 J. F. Johnson and E. M. Barral, II, J. Chromatogr., 31 (1967) 547.
142 O. L. Hollis and W. V. Hayes, J. Gas Chromatogr. 4 (1966) 235.
143 R. M. Cassidy, M. T. Burteau, J. P. Misian and R. W. Ashley, J. Chromatogr. Sci., 14 (1976) 444.
144 R. L. Malcolm, E. M. Thurman and G. R. Aiken, Proceedings of the 11th Annual Conference on Trace Substances in Environmental Health, Columbus, MO, 1977, pp. 307-314.
145 P. van Rossum and R. G. Webb, J. Chromatogr., 150 (1978) 381.
146 P. R. Musty and G. Nickless, J. Chromatogr., 89 (1974) 185.
147 S. F. Stepan and J. F. Smith, Water Res., 11 (1977) 339.
148 C. D. Chriswell, R. L. Ericson, G. A. Junk, K. W. Lee, J. S. Fritz and H. J. Svec, J. Amer. Water Works Ass., 69 (1977) 669.
149 G. Marcelin, J. Chromatogr., 174 (1979) 208.
150 J. Solomon, Anal. Chem., 51 (1979) 1861.
151 R. Baird, M. Selma, J. Haskins and D. Chappelle, Water Res., 13 (1979) 493.
152 R. C. Chang and J. S. Fritz, Talanta, 25 (1978) 659.
153 L. R. Snyder and J. J. Kirkland, Introduction to Modern Liquid Chromatography, Wiley, New York, 2nd ed., 1979, p. 568.
154 K. Blau and G. S. King, Handbook of Derivatives for Chromatography, Heyden, London, Bellmawr, NJ, 1977.
155 J. F. Lawrence and R. W. Frei, Chemical Derivatization in Liquid Chromatography,

Elsevier, Amsterdam, 1976.
156 J. F. Lawrence (Ed.), Special Issue, Derivatization in Chromatography, J. Chromatogr. Sci., 17 (1979) 113-176.
157 M. R. Guerin, G. Olerich and W. T. Rainey, Anal. Chem., 56 (1974) 761.
158 A. E. Pierce, Silylation of Organic Compounds, Pierce Chemical Company, Rockford, IL, 1968.
159 P. S. Mason and E. D. Smith, J. Gas Chromatogr., 4 (1966) 398.
160 T. Walle and H. Ehrsson, Acta Pharm. Suec., 7 (1970) 389.
161 F. Karoum, C. F. R. Ruthven and M. Sandler, Biochem. Med., 5 (1971) 505.
162 J. B. Brooks, J. A. Liddle and C. C. Alley, Anal. Chem., 47 (1975) 1960.
163 T. H. Jupille, Amer. Lab., May (1976) 85.
164 H. D. Durst, M. Milano, E. J. Kikta, S. A. Connelly and E. Grushka, Anal. Chem., 47 (1975) 1797.
165 R. W. Roos, J. Chromatogr. Sci., 14 (1976) 505.
166 J. W. Higgins, J. Chromatogr., 121 (1976) 329.
167 E. Nachtmann and K. W. Budna, J. Chromatogr., 136 (1977) 279.
168 J. W. Vogh, Anal. Chem., 43 (1971) 1618.
169 H. M. Fales and T. Lauakkainen, Anal. Chem. 37 (1965) 955.
170 J. Korolczuk, M. Daniewski and Z. Mielniczuk, J. Chromatogr., 88 (1974) 177.
171 L. J. Papa and L. P. Turner, J. Chromatogr. Sci., 10 (1972) 744.
172 H. Kallio, R. R. Linko and J. Kaitatanta, J. Chromatogr., 65 (1972) 355.
173 R. J. Soukup, R. J. Scarpellino and E. Danielczik, Anal. Chem., 36 (1964) 2255.
174 F. A. Fitzpatrick, M. A. Wynalda and D. G. Kaiser, Anal. Chem., 49 (1977) 1032.
175 N. Narasimhachari and P. Vouros, Anal. Biochem., 45 (1972) 154.
176 S. Baba, J. Chromatogr., 88 (1974) 373.
177 T. Walle, Acta Pharm. Suec, 5 (1968) 367.
178 D. J. Edwards and K. Blau, Anal. Biochem., 45 (1972) 387.
179 D. R. Knapp and S. Krueger, Anal. Lett. 8 (1975) 603.
180 H. D. Durst, M. Milano, E. S. Kikta, S. A. Connelly and E. Grushka, Anal. Chem., 47 (1975) 1797.
181 "Fluorotags", Technical Bulletin, Regis Chemical Co., Morton Grove, IL.
182 J. P. Thenot and E. C. Horning, Anal. Lett. 5, (1972) 519.
183 M. G. Horning, A. M. Moss and E. C. Horning, Biochem. Biophys. Acta, 148 (1967) 597.
184 E. Anggard and G. Sedvall, Anal. Chem., 41 (1969) 1250.
185 M. Donike, J. Chromatogr., 78 (1973) 273.
186 P. Cancalon and J. D. Klingman, J. Chromatogr. Sci., 10 (1972) 253.
187 M. Donike, J. Chromatogr., 103 (1974) 91.
188 M. Donike, Chromatographia, 7 (1974) 651.
189 S. B. Matin and M. Worland, J. Pharm. Sci., 61 (1973) 1235.
190 M. G. Horning, A. M. Moss, E. A. Boucher and E. C. Horning, Anal. Lett., 1 (1968) 311.
191 J. Vessman, K. Hartvig and M. Holander, Anal. Lett. 6 (1973) 699.
192 B. Maume, P. Bournot, J. C. Lhugenot, C. Baron, F. Barbier, G. Maume, M. Prost and P. Padieu, Anal. Chem., 45 (1973) 1073.
193 D. D. Clarke, S. Wilk and S. E. Gitlow in H. A. Szymanski (Ed.) Biomedical Applications of Gas Chromatography, Plenum Press, New York, 1964, p. 53.
194 J. C. Lhugenot and B. F. Maume, J. Chromatogr. Sci., 12 (1974) 411.
195 F. Karoum, F. Cattabeni, E. Costa, C. R. J. Ruthven and M. Sandler, Anal. Biochem., 47 (1972) 550.
196 E. Gelpi, E. Peralta and J. Segura, J. Chromatogr. Sci., 12 (1974) 701.
197 E. C. Horning, M. A. Horning, W. J. A. VandenHeuvel, B. Holmstedt and C. J. W. Brooks, Anal. Chem., 36 (1964) 1546.
198 D. D. Clarke, S. Wilk and E. Gitlow, J. Gas Chromatog., 4 (1966) 310.
199 P. Hartvig and J. Vessman, Anal. Lett. 7 (1974) 223.
200 P. Hartvig and J. Vessman, J. Chromatogr. Sci., 12 (1974) 722.
201 P. Hartvig, W. Handl, J. Vessman and C. M. Smith, Anal. Chem., 48 (1976) 390.
202 H. V. Street, J. Chromatogr., 73 (1972) 73.
203 C. R. Clark and M. M. Wells, J. Chromatogr. Sci., 16 (1978) 332.
204 F. S. Tanaka and R. G. Wien, J. Chromatogr., 87 (1973) 85.
205 K. L. Dunlap, R. L. Sandridge and J. Keller, Anal. Chem., 48 (1976) 497.

206 S. J. Yu, U. Kiigemazi and L. C. Teriere, J. Agr. Food, 19 (1971) 5.
207 G. T. Flint and W. A. Aue, J. Chromatogr., 52 (1970) 478.
208 I. E. Smiley and E. D. Schall, J. Ass. Offic. Anal. Chem., 52 (1969) 107.
209 N. E. Hoffman and K. A. Peteranetz, Anal. Lett., 5 (1972) 589.
210 F. K. Kawahara, Anal. Chem., 40 (1968) 1009.
211 I. C. Cohen, J. Norcup, J. H. A. Ruzicka and B. B. Wheals, J. Chromatogr., 44 (1969) 251.
212 H. Ehrsson, J. T. Walle and H. Brotell, Acta Pharm. Suec., 8 (1971) 319.
213 J. Muth and J. Giles, Alter Chromatogram, 3, No. 2 (1980) 5.
214 R.-D. Penzhorn and W. G. Filby, Staub-Reinhalt Luft, 36 (1976) 205; Anal. Abstr., 33 (1977) 2H14.
215 L. L. Lamparskii and T. J. Nestrick, J. Chromatogr., 156 (1978) 143.
216 C. S. Giam, H. S. Chan, T. F. Hammargren, G. S. Ness and D. L. Stalling, Anal. Chem., 48 (1978) 78.
217 W. P. Cochrane, J. Chromatogr. Sci., 17 (1979) 124.
218 I. Sunshine, Methodology for Analytical Toxicology, CRC Press, West Palm Beach, FL, 1975, p. 116.
219 T. D. Luckey and R. Venugapol, Metal Toxicity in Mammals, Plenum Press, New York, 1977, Ch. 1 and 5.

RELATED READINGS ON SAMPLE TREATMENT

1 The Gas Phase Extraction and Analysis of Volatile Organic Pollutants In Water Using a Tenax-GC Adsorption Tube. B.I. Brookes, S.M. Jickells and R.S. Nicolson, J. Assoc. Public Anal., 16(4)(1978)117-21.

2 Problems and Pitfalls in Acetaldehyde Determinations. C.J.P. Eriksson, Alcohol. Clin. Exp. Res., 4(1)(1980)22-9.

3 Tenax-GC Extraction Technique for Residual Polychlorinated Biphenyl and Polyaromatic Hydrocarbon Analysis in Biodegradation Assays. M.P. Shiaris, T. Sherrill and G.S. Sayler, Appl. Environ. Microbiol., 39(1)(1980)165-71.

4 Study of the Composition and Content of Petroleum Hydrocarbons in Baltic Sea Water Using Capillary Gas Chromatography and IR Spectrophotometry. K. Tiks, Sint. Issled. Biol. Swedin., Tezisy Dokl. Konf. Molodykh Uch., 6th(1978)91-2.

5 Determination of Petroleum Hydrocarbons in Tap Water by Gas Chromatography. V. Durdovic, Acta Hydrochim. Hydrobiol., 7(5)(1979)527.

6 Gas Chromatographic Headspace Analysis in Brewing. E. Geiger. In Applied Headspace Gas Chromatography, B. Kolb, ed., Heyden, London, 1980, pp. 73-9.

7 GC Headspace Analysis Used in Government Foodstuffs Control-Applications for Testing Requisites. J.J. Doemling. In Applied Headspace Gas Chromatography, B. Kolb, ed., Heyden, London, 1980, pp. 171-8.

8 Simple Rapid and Sensitive Method for the Simultaneous Quantitation of Ethanol and Acetaldehyde in Biological Materials Using Head-Space Gas Chromatography. C.L. Mendenhall, J. MacGee, and E.S. Green, J. Chromatogr., 190(1)(1980)197-200.

9 Analysis of Headspace Gases for Parts Per Billion Concentrations of Volatile Organic Contaminants in Water Samples by Gas Chromatography. D.A.J. Murray, Environ. Sci. Res., (1980)207-16.

10 Measurement and Analysis of Air Pollution. 6. Determination of Trimethylamine in Air. S. Sano, Kanagawa-ken Taiki Osen Chosa Kenkyu Hokoku, 21(1979)117-23.

11 Acetaldehyde, Methanol, and Ethanol Analysis by Headspace Gas Chromatography. R.M. Anthony, C.A. Sutheimer and I. Sunshine, J. Anal. Toxicol.,4(1)(1980) 43-5.

12 Determination of Residual Acetaldehyde in Poly(ethylene terephthalate) Bottles, Preforms and Resins by Automated Headspace Gas Chromatography. M. Dong, A.H. DiEdwardo and F. Zitomer, J. Chromatog. Sci., 18(5)(1980)242-6.

13 A Gas Chromatographic Method for the Accurate Determination of Low Concentrations of Volatile Sulfur Compounds in Alcoholic Beverages. O.Leppanen, J. Denslow and P. Ronkainen, J. Inst. Brew., 85(6)(1979)350-3.

14 Gas Chromatographic Head-Space Assay of Formic Acid as Methyl Formate in Biologic Fluids: Potential Application to Methanol Poisoning. C.Abolin, J.D. McRae, T.N. Tozer and S. Takki, Biochem. Med. 23(2)(1980)209-18.

15 Determination of Hydrogen Cyanide in Blood Using Gas Chromatography with Alkali Thermionic Detection. R.W. Darr, T.L. Capson and F.D. Hileman, Anal. Chem., 52(8)(1980)1379-81.

16 Supplementary Report to"Gas Chromatographic Determination of COHb by the Use of a Heated Gas Sampler." K. Yamamoto, and Y. Yamamoto, Nippon Hoigaku Zasshi, 33(6)(1979)718-20.

17 Study of Flavor of Wines from "Tierra de Barros" (Spain) by Phase Separation Through Salt Addition and Gas Liquid Chromatography of the Organic Phase. J.L. Mesias, J.I. Maynar and I. Mareca, Rev. Agroquim. Tecnol, Aliment, 20(2)(1980)240-6.

18 Methods of Derivatization in Analytical Gas Chromatography: General Principles and Recent Developments. A. Pannatier, B. Testa, Pharm. Acta Helv., 55(4)(1980)100-13.

19 Use of Headspace Gas Chromatographic Analysis for Determination of Volatile Impurities in Polymers. B.V. Ioffe and T.L. Reznik, Zh. Anal. Khim., 35(7) (1980)1410-27.

20 Enhancement of Electron-Capture Detection of Chlorocarbons by Iodination. A.J. Watson, G.L. Ball and D.H. Stedman, Anal. Chem., 53(1)(1981)132-4.
21 Isolation and Characterization of Odorous Components in Solid Swine Manure. A. Yasuhara, and K. Fuwa, Agric. Biol. Chem., 44(10)(1980) 2379-85.
22 Determination of Halogenated Hydrocarbons in Silicon Halides with Two Gas-Chromatographic Methods and a Comparison of These Analyses.H.J. Rath, and J. Wimmer, Fresenius' Z. Anal. Chem., 303(1)(1980)14-17.
23 Determination of Cyanides and Thiocyanates in Water by Headspace Gas Chromatography with a Nitrogen-Phosphorus Detector. G. Nota, V.R. Miraglia, C. Improta and A. Acampora, J. Chromatogr., 207(1)(1981)47-54.
24 Application of Headspace Gas Chromatography to the Determination of Chlorinated Hydrocarbons in Waste Waters. L. Lukacovic, M. Mikulas, A. Vanko and G. Kiss, J. Chromatogr., 207(3)(1981)373-7.
25 Preconcentration for Trace Analysis of Organic Compounds. F.W. Karasek, R.E. Clement and J.A. Sweetman, Anal. Chem., 53(1981)1050 A.
26 Preconcentration for Trace Element Determination in Aqueous Samples. D. E. Leyden and W. Wegscheider, Anal. Chem., 53(1981)1059A.

CHAPTER 5

THE USE OF GAS CHROMATOGRAPHY IN ENVIRONMENTAL ANALYSES

5.1 INTRODUCTION

The preceding chapters have discussed the environmental problem in general (Chapter 1), the criteria used for sampling (Chapter 2), the sampling process (Chapter 3) and the sample treatment (Chapter 4). We can now discuss the analyses of these environmental samples, using gas chromatography and liquid chromatography (Chapters 5 and 6).

There is a vast amount of scientific literature available regarding the use of gas chromatography for environmental analyses, much more than exists for liquid chromatography. There are a number of reasons for this; paramount among these is the fact that gas chromatography has been more readily available when environmental problems came to the forefront and that the majority of compounds to be determined are easily handled by the gas chromatographic process. This is especially true when an air pollution problem exists. To present an all-inclusive discussion of analyses by gas chromatography would result in a very long monograph itself. Thus, we shall highlight this area of analysis and furnish many easily obtainable references. To condense many of these references we will make use of tables to summarize the salient features of the articles.

5.2 STANDARDS AND CALIBRATION

We have presented in Chapter 2 sources of many reference (standards) components which are available. Calibration by use of these standard components is a necessary procedure to have quantitative information for any environmental sample system. Meaningful information will result if reliable information is given regarding the volume of sample, volume of standard solution, and flow measurements for gas and liquid samples.

Perhaps one of the most difficult aspects of calibration and quantification is being able to express the small concentrations found in environmental samples. Table 5.1 summarizes many of the mass-volume or volume-volume relations encountered in the area of the analytical chemistry of environmental systems.

The preparation of standard aqueous solutions is best performed by starting with a concentrated solution and making the appropriate dilutions to furnish less concentrated standards. The choice of containers for these solutions should be predicated upon the theory that loss of solute and contamination from the container must be minimized. These two goals can be achieved by using clean glass volumetric flasks and/or vials with either standard tapered stoppers or screw-on lids lined with Teflon (PTFE) or aluminum foil. Once prepared, they should be stored in a refrigerator or freezer. Very dilute standard solutions (10 ppm or less) are best prepared when needed. Prolonged storage of very dilute standards raises questions regarding their stability. Adsorption on walls of containers and/or loss by volatilization are the most frequent causes of this decreased concentration.

TABLE 5.1
Mass-volume and volume-volume relationships for environmental samples.

I. Mass-volume

mg/ℓ	=	1 ppm	=	1000 ppb
μg/ml	=	1 ppm	=	1000 ppb
μg/5.0 ml	=	0.2 ppm	=	200 ppb
μg/10.0 ml	=	0.10 ppm	=	100 ppb
μg/20.0 ml	=	0.05 ppm	=	50 ppb
μg/100.0 ml	=	0.01 ppm	=	10 ppb

Equation for calculation: (grams/1000 ml) × 10^6 = ppm
mg/ml = 1000 ppm = 1,000,000 ppb

II. Volume-volume

μl/ml	**= 1000 ppm = 10^6 ppb = 0.1%**
μl/10.0 ml	**= 100 ppm = 10^5 ppb = 0.01%**
μl/100.0 ml	**= 10 ppm = 10^4 ppb = 0.001%**
μl/1000.0 ml	**= 1 ppm = 10^3 ppb = 0.0001%**
100 μl/10.0 ml	**= 10,000 ppm = 10^7 ppb = 1.0%**
10 μl/1.0 ml	**= 10,000 ppm = 10^7 ppb = 1.0%**

Equation for calculation: (μl/ml) × 10^6 = ppm

III. Inter-relationships

%	ppm	ppb	mg/100 ml	g/μl	UNIT
100.00	1,000,000		100,000		
10.00	100,000		10,000		
1.00	10,000		1,000		
0.1	1,000		100	10-6	μg
0.01	100		10	10-7	
0.001	10		1	10-8	
0.0001	1	1,000	0.1	10-9	ng
0.00001	0.1	100	0.01	10-10	
0.000001	0.01	10	0.001	10-11	
0.0000001	0.001	1	0.0001	10-12	pg

Preparation of aqueous standards presents another problem, that of "pure" water solvent. This "pure" water must be free of soluble organic substances (volatile as well as non-volatile). In many cases, regular distilled water is not adequate and must be purified further.

(1) Volatile contaminants. If only volatile contaminants are present, the "pure" water may be prepared by boiling the water in a clean atmosphere. Standards prepared in "pure" distilled water should have no gas phase above the liquid phase. The standard solutions should completely fill the container.

(2) Non-volatile contaminants. These contaminants may be removed by using the sampling techniques discussed in Chapter 3. Adsorbents used can be Porapak Q or Bondapak (ref. 1), Amberlite XAD-2 resin or activated charcoal (ref. 2 and 3) and Tenax GC (ref. 4). Creed (ref. 1) cautions that "pure" water made by these techniques should not be stored longer than three days. After the first couple of days, the blank correction begins to increase.

(3) Non-aqueous solvents. It has been the authors' experience that large volumes of "clean" solvents (> 500 ml) should be stored in another room than the laboratory. Only small volumes (ca. 100 ml) should be kept in the laboratory. Refilling of the laboratory supply should be performed in the storage area. A good philosophy for solvents used in trace environmental analysis is that no solvent should be considered pure. Acetone and methanol can be purified by distillation in a glass apparatus, whereas n-pentane, n-hexane and isooctane should be distilled after sulfonation of the aromatic impurities and nitration (ref. 5).

Restrictions placed on solvents and containers are more stringent than those placed on the purity of the standards. Standard compounds need be no higher purity than 99% for environmental trace analysis.

Solubility of many organic compounds (e.g., hydrocarbons) in water is very low and aqueous standard solutions may have to be prepared by using an organic solvent which is soluble in water. A disadvantage to this technique is that, in subsequent extraction of the aqueous solution, a large quantity of the organic solvent is carried through and may overlap solute peaks of interest in the chromatogram.

Air or gas samples offer a unique application for gas chromatography. The solvent here is air and blanks must be made from "zero air." The term "zero air" appeared in Title 40 of the U.S. Federal Regulations, Part 86 of Control of Air Pollution for New Motor Vehicles and New Motor Vehicle Engines, which had a requirement for the flame ionization detector (FID) and other analyzers. Specifications for "zero air" are shown in Table 5.2.

TABLE 5.2

Zero air specifications from U.S. Federal Register 33(108), (1968), Part II

< 1 ppm (v/v)* total hydrocarbons as carbon
< 1 ppm (v/v)* carbon monoxide
< 400 ppm (v/v)* carbon dioxide
< 0.1 ppm (v/v)* nitric oxide
18.0 - 21.0 (v/v)* oxygen

*See Table 5.1 for explanation of concentrations.

The laboratory air and outside air may meet these criteria, and no purification procedure is necessary. Preparation of "zero air" may be produced in a number of ways (ref. 6).

(1) Purification of the normal atmospheric air on site;
(2) Purification of normal purity oxygen and nitrogen, then blended in proper ratio;
(3) Purchasing special purity oxygen and nitrogen which is then mixed in the gas phase for proper ratio;
(4) Purchasing "zero air" commercially.

One defines sensitivity of an analytical technique as the lowest concentration (smallest amount) of a component that can be determined with certainty. To be certain that the concentration measured is meaningful, one must determine a blank or zero air for air samples. The blank is the same volume as the sample but does not contain the pollutant in question. "Zero air" is synonomous with "reagent blank" in other analytical techniques. Usually the blank is run concurrently with the samples. Over a period of time, one obtains fluctuations in the value of the blank. The arithmetic mean of these blanks can be used as the "zero value."

Many analytical data terms are used loosely. Perhaps the two most common terms which are interchanged are accuracy and precision (see Chapter 3). Reproducibility of a method is a measure of the repeatability (precision) but accuracy indicates the error of the mean, i.e., how close the experimental value agrees with the "true value." It is not always necessary that a method possesses both qualities; the monitoring of trends in a particular air or water pollution study requires a reproducible analytical method more than an accurate analytical method.

Calibration and standardization are often used incorrectly. We will use the word calibration to mean comparing an instrument or analytical technique to a reference that is constant, but not necessarily representative of the "true" concentration. Standardization implies that there is a one-to-one relationship between the analytical technique (or instrument) and the "true" concentration. Accurately-prepared test atmospheres are used for both calibration and standardization.

The defining of analytical systems for environmental work may be carried one step further to set forth clarity of terms. Primary systems are those in which the concentrations of trace substances are calculated from flow rates and physical constants. Secondary systems are those in which the trace substance concentration is measured as it appears in the system.

Meaningful data are obtained from environmental samples only when comparison is made to some standard. Thus, all equipment used from the sampling step to the quantitation step must be calibrated. The greatest errors occur either when concentration levels being measured are very low (< ppb) or when chemisorption effects are significant. Solid or liquid standards have a longer stability than do gaseous standards; this is especially true at low concentrations (< ppm). Standards for the ppb range should be prepared as needed. Changes in concentration of standards usually are due to sorption effects at the walls of the container, permeation through the container walls and/or photochemical effects. There is no one calibration or standardization technique which is workable for all types of samples; the analytical chemist must evaluate the problem from the aspects of sampling through the separation and quantitation steps. The best conditions are those which are performed with the minimum of effort and yield the maximum of reliable information. A parameter that is often overlooked is the detector response. It makes little difference if a thermal conductivity detector (TCD), a flame ionization detector (FID), an electron capture detector (ECD), or a mass spectrometer detector (MSD) is used. Each is affected by the operating parameters and the type of sample being separated.

The use of standards and/or calibration techniques requires an understanding of the system and the instrumentation required for the analysis. As a rule of thumb, one should prepare between 5 to 10 times the volume of standard required for one measurement.

Some of the more commonly employed calibration techniques are: dynamic calibration gas generator (refs. 7-10), diffusion cells (refs. 11-13), dilution (ref. 14-19), permeation tubes (refs. 13 and 20-24), exponential dilution flask (refs. 13, 25 and 26), rotating syringe technique (ref. 27), static systems for producing gas mixtures (ref. 10), preparation of weight/weight, weight/volume, or volume/volume solutions (see Tables 5.1, 5.3 and 5.4), and plastic bag calibration techniques (refs. 15, 28 and 29).

5.2.1 Volume measurement and standards for air samples

Data for air pollutants are usually referred to a m^3 of air and the amount of pollutant expressed in terms of weight (mg/m^3 or $\mu g/m^3$) or as volume ratio ($1:10^6$ = parts per million = ppm or $1:10^9$ = parts per billion = ppb). Ratio expressions (ppm or ppb) are independent of pressure and temperature, whereas weight/volume (mg/m^3 or $\mu g/m^3$) expressions vary according to the gas laws. In the Federal Republic of Germany, pollutant concentrations are expressed as mg/1 Nm^3; meaning 1 m^3 of dry gas at Normal Temperature and Pressure (ref. 30).

Weight/volume values (mg/1 Nm^3 may be converted to ratio expressions (ppm or ppb) by the following equation:

$$\text{ppm} = (\text{mg}/1\ \text{Nm}^3)(\text{molar volume}/\text{molecular weight}) \qquad (5.1)$$

Table 5.3 summarizes the more commonly used relationships and calculations for gaseous systems. This table complements Table 5.1 since it presents specific calculations and relationships for gaseous systems.

Two parameters are necessary for the preparation of standard gas mixtures. These are the flow rate and volume measurements. Calibration procedures are undertaken to relate detector response to concentration of component being determined. Another justification for this procedure is that the correlation between detector response (integral response) and sample concentration (weight or volume relations) is not linear over the entire concentration range used.

TABLE 5.3

Measurements and calculations of concentrations

Volume/volume relations

Concentration (ppm) = $[v_a/(V_D + v_a)] \times 10^6$ (5.2)

v_a = volume of solute; V_D = volume of solvent (diluent gas)

Concentration (ppm) = $[p_a/(P_D + p_a)] \times 10^6$ (5.3)

p_a = partial pressure solute; P_D = partial pressure of solvent (diluent gas).

At concentrations < 5000 ppm, the v_a or p_a terms in the denominators may be dropped. This causes < 0.5% relative error.

ppm = micromoles of solute gas/mole of solvent gas

ppm $\times$ (1×10^{-4}) = %(v/v)

ppm (mol. wt./24.45) = mg/m^3; where, 24.45 = (62.361 $\times$ 298°K)/760 Torr = RT/P.

1 ppb $\times$ (1×10^{-3}) = 1 ppm

1% (v/v)(10,000) = 1 ppm

(mg/ℓ)(1000) = mg/m^3

(mg/ℓ)(1×10^6) = µg/m^3

Concentration (ppm) = [(mg/m^3 $\times$ 24.450 $\times$ T°K $\times$ 760 Torr)/(mol. wt. $\times$ 298°K $\times$ P(Torr)] (5.4)

Concentration (mg/m^3) = [(conc.(ppm) $\times$ mol. wt. $\times$ 298°K $\times$ P(Torr))/(24.454 $\times$ T°K $\times$ 760 Torr)] (5.5)

Weight/volume relations

Concentration (w/v) = W_a/V_D (5.6)

W_a = weight of solute (mg); V_D = volume solvent (diluent gas)

Concentration in w/v usually given in mg/m^3

1m^3 = 10^6 ml; 10^3 ml = 1 liter;

thus 10^6 ml = 1000 liters or 1 m^3 = 1×10^9 mm^3 = 264.17 U.S. gallons

TABLE 5.3 (Cont.)

$$\text{concentration (ppm)} = (W_a/V_D) \times 10^6 \tag{5.7}$$

$n_a = W_a/(MW)_a$ and $PV_a = n_aRT$; thus

$PV_a = (W_a/(MW)_a)RT$ P = barometric pressure

$W_a/V_a = P(MW)_a/RT$ or $V_a = W_aRT/(MW)_aP = v_a$

$\text{Concentration (ppm)} = (v_a/V_D)10^6$ from (5.2)

$$= ((W_aRT)/(MW_aPV_D)) \times 10^6 \tag{5.8}$$

$$\text{Concentration (ppm)} = ((\rho_a v_a RT)/((MW)_aPV_D)) \times 10^6 \tag{5.9}$$

ρ_a = density = W_a/v_a

Relationship of concentration (w/v) and concentration (ppm)

Using eqns 5.6 and 5.8; solve eqn 5.8 for W_a and substitute into eqn 5.6

$W_a = \text{Conc. (ppm)}(MW)_aPV_D/10^6RT$

$\text{Conc. (w/v)} = (\text{Conc. (ppm)}(MW)_aPV_D/10^6RT)/V_D$

$$= (\text{Conc. (ppm)} \times 10^{-6}(MW)_aP)/RT \tag{5.10}$$

In terms of vapor pressure:

$\text{Concentration w/v} = ((MW)_aP/RT)(P_a/P_D)$

At STP: $P_D \cong P$

$$\text{Thus, concentration (w/v)} = (MW)_aP_a/RT \tag{5.11}$$

Common conversion units

Dynes/cm^2 × (1.4504 × 10^{-5}) = pounds/inch^2 = psi
Dynes/cm^2 × (10.197 × 10^{-4}) = grams/cm^2(absolute)
Pounds/in^2(absolute) × 70.307 = grams/cm^2(absolute)
Grams/cm^3 × 1 = grams/ml (at 4°C only)
Grams/cm^3 × 0.03613 = pounds/inch^3
Grams/cm^3 × 8.3452 = pounds/U.S. gallons
Grams/cm^3 × 62.428 = pounds/ft^3

TABLE 5.3 (Cont.)

Pounds/ft^3 x 0.0602 = grams/cm^3

Pounds/ft^3 x (5.7870 x 10^{-4}) = pounds/inch^3

Pounds/in^2 x 6893 = Pa

ppm (v/v) x (1 x 10^{-6}) = partial pressure of one constituent divided by total pressure of mixture

mg/m^3 x (1 x 10^{-3}) = mg/ℓ

μg/m^3 x (1 x 10^{-6}) = mg/ℓ

mg/ℓ x (1 x 10^3) = mg/m^3

mg/ℓ x (1 x 10^6) = μg/m^3

μg/m^3 x (1 x 10^{-6}) = mg/ℓ

Angstrom units x (1 x 10^{-10}) = m

Angstrom units x (1 x 10^{-4}) = micrometers, μm

Angstrom units x (1 x 10^{-8}) = cm

μm x (1 x 10^{-6}) = m

mm x 1000 = μm

cm x (1 x 10^4) = μm

mm^3 x (1 x 10^{-9}) = m^3

m^3 x 264.17 = U.S. gallons

ft^3 x 0.02832 = m^3

ft^3 x 28.316 = ℓ

cm^3 x (1 x 10^{-6}) = m^3

(i) The dynamic calibration technique is useful when one needs a continuous source of a calibrated gas or when the gases under study are unstable. The concentrations are readily calculated from either the vapor pressure of the liquids or the flow rates of the liquids or the flow rates of the gases employed. It has been found that the equipment used is slow to equilibrate (at low concentrations) and thus, changing from one sample type to another is time consuming. It is possible to prepare standards in which the concentration of one component ranges from several ppm to 50%. The technique is not useful for compounds which are liquids at temperatures lower than 0°C.

Each contaminant gas concentration can be varied by changing its flow rate in the diluent stream. Concentration in the gas stream is based upon the partial flow rates of each gas:

$$C_{cg} = ((F_{cg})/(F_{cg} \times F_{dg})) \times 10^6 \qquad (5.12)$$

where: C_{cg} = volume concentration of contaminant gas in ppm; and F_{cg} and F_{dg} = flow-rates of contaminant and diluent gases, respectively.

Angely et al. (ref. 31) have published a technique for the preparation of standard samples for detector calibration, whereas Averett (ref. 32) designed an all-glass apparatus suitable for preparing corrosive-gas standards; pressures were monitored by a mercury manometer. Brockway et al. (ref. 33) have described a method for the simultaneous calibration of gas analyzers and meters. Axelrod et al. (ref. 34) made a pressure device of ceramic to prepare gas contaminants at the 100 ppm level. The ceramic disk was connected directly to a high pressure tank of the contaminant gas, and the flow rate through the ceramic frit precisely controlled by regulation of the pressure.

A novel method of generation of gases at the ppm level was developed by Hashimoto and Tanaka (ref. 35) which enables the generation of such gases as SO_2, NO_x(NO and NO_2), HCN, NH_3 and H_2S by a simple apparatus and simple operations. The generated gases can be utilized for calibration of measuring instruments, attainment of biological and medical tests and the study of chemical reactions in ambient air. The solution for gas generation contains a specific solute for each gas: e.g., sodium hydrogen sulfite for SO_2, sodium nitrate for NO_x, sodium sulfide for H_2S, ammonium chloride for NH_3 and potassium cyanide for HCN. The pH of the generating solution is controlled by buffering at a specific pH. The pH range of the solution for generation of gases in the range of 0.01-0.1 ppm for SO_2 is 5.0-6.0, for NO_x 6.0-7.0 and for H_2S 12-13.

A linear logarithmic relationship was found between concentration in the gas phase, C_{gas}(in ppm) and concentration in solution, C_{soln}(in µg/ml). It was thus presumed that the gas generation occurred by diffusion due to concentration differences expressed in Fick's law.

Hashimoto and Tanaka state that the advantages of their method over conventional permeation tubes are: (1) no requirement for tube preparation, (2) smaller temperature dependence, and (3) faster equilibration. Other gases, e.g., CO, CO_2 and HF, can be generated by this method.

(ii) The operation of permeation tubes is based on the fact that nearly all plastic materials will retain a liquid while allowing its vapor to dissolve in the walls and diffuse at a calculable rate. The diffusion rate is an exponential function of absolute temperature and a direct function of wall thickness and the surface area of the tube. The rate of diffusion is also dependent upon the molecular weight of any carrier (diluent) gas which passes over the outside wall and its moisture content (if the permeating gas reacts with water).

Fick's law of diffusion approximately describes the diffusion rate of gases from the permeation tubes:

$$D = dsa(P_i - P_o)/W \tag{5.13}$$

where: D = volume of the diffusing gas;
d = the gas diffusion constant;

s = solubility of gas in tube material;
a = area of the tube;
W = tube wall thickness;
P_i = inside pressure of tube;
P_o = outside pressure of tube.

To assure constant permeation rate, constant temperature is maintained (solubility(s) and diffusion rate(s) are temperature-dependent). The attainment of a permeation rate is not instantaneous; thus one allows the tube and the system to come to equilibrium before actually performing the calibration.

Permeation rates are calculated by measuring the weight loss (using a microbalance) with time. Weight losses are usually expressed as nanograms per centimeter of tube length per minute (ng/cm. min). O'Keefe and Ortman (ref. 20) have presented a detailed study of permeation tube calibration. In this study both the inside diameter and wall thickness of the tubes were varied for such gases as SO_2, NO_2, C_3H_8, CO_2, C_6H_6, etc.

The tube length, L, necessary for a particular calibration, may be calculated by dividing the desired rate, R_R, of permeation (ng/min) by the rate, R_M, of a tube of similar material in ng/cm. min., i.e.,

$$L = R_R/R_M \qquad (5.14)$$

An outstanding feature of permeation tubes is that the permeation rates remain relatively constant over the life of the tubes.

(iii) McKelvey and Hoelscher (ref. 11) reported the first diffusion cells which allowed easy preparation of low concentrations of volatile materials in air. The cell consisted of two flasks joined by a diffusion tube. The lower tube contained the liquid and saturated vapor. This saturated vapor would diffuse to the top flask where it would mix with a flowing diluent gas. The rate of diffusion, γ, could be calculated:

$$\gamma = 2.303(DPMA/RTL)\log P/(P-p) \qquad (5.15)$$

where: D = diffusion coefficient of the vapor;
P = total pressure in chambers (atm);
M = molecular weight of diffusing liquid vapor;
A = cross-sectional area of connecting diffusion tube (cm^2);
R = gas constant (liter-atm/mol°K);
T = absolute temperature (°K);
L = length of diffusion tube (cm);
P = vapor pressure of diffusion liquid vapor at T(atm).

(iv) The exponential dilution flask offers a unique way of preparing standards for calibration. With this technique one can prepare a dynamic range of concentrations of the contaminant gas. The diluent gas enters the top of the flask, is mixed with an initial known concentration of gas (by means of a magnetic stirrer or paddle wheel assembly), then exits at the bottom of the flask. Knowing the gas flow rate of the diluent gas and volume of the mixing chamber, one can calculate the concentration, C, of contaminant gas at any time:

$$C = C_o e^{-tQ/V} \quad (5.16)$$

where: C_o = initial gas composition;
V = volume of flask;
Q = the flow rate;
t = time.

One outstanding feature of this technique is that it allows one to study the dynamic range of a gas chromatographic detector in a simple experiment.

Bruner et al. (ref. 26) and Folmer (ref. 36) have combined permeation tubes and the exponential dilution flask. The disadvantage of this combination is that the calibration curve begins to lose linearity as the contaminant gas approaches very low concentrations (which may be attributed to adsorption of the contaminant gas on the walls of the container). A variation of this type of calibration is the syringe-dilution technique where successive dilutions of the gas sample are made with a glass injection syringe (ref. 36).

5.2.2. Volume measurement and standards for water samples

Measurement of water samples can involve any of the three states of matter (gas, liquid or solid) dissolved in the aqueous solvent or any combination(s) of the three. The treatment of gases, in solution, has been treated in Chapter 4 (Section 4.2.2.1 Headspace and/or vapor equilibration) and will not be discussed in this section.

Water samples are usually taken in a fixed volume container and an aliquot is then analyzed in the laboratory. Knowing the volume of the aliquot, the volume of water brought to the laboratory and the total volume of the water source or its flow rate at the sampling site, one may relate the analysis data to the contaminant(s) concentration(s) in the water supply (see Chapter 3 for additional information concerning sampling of liquids).

Table 5.4 illustrates many of the most common terms and equations for expressing measurements and concentration calculations for liquid samples.

5.3 SAMPLE INTRODUCTION ONTO THE COLUMN

The gas chromatographic system is necessary for the transferring of a measured amount of the sample material onto the column. A sample size of 0.001 ml (1 μl) to 10 ml may be injected into the carrier gas stream to obtain quantitative and reproducible data (the smaller volumes are used for liquid samples whereas larger sample volumes may be used for gaseous samples). Maximized chromatographic response requires that the sample be contained in the smallest column volume possible. The

TABLE 5.4

Measurements and calculations of concentration

Liquids in liquids

Concentration (ppm) = $(v_a/(V_D + v_a))10^6$

v_a = volume of solute and V_D = volume of solvent (diluent).

At concentration < 500 ppm, the v_a term in the denominator may be dropped. This gives a maximum of 5.0% relative error.

Example:

100 µl/10 ml

$$\begin{aligned} \text{conc.(ppm)} &= 100\ \mu l/(10\ ml + 100\ \mu l)) \times 10^6 \\ &= (0.1\ ml/(10\ ml + 0.1\ ml)) \times 10^6 \\ &= (0.1/10.1) \times 10^6 = 9.9 \times 10^{-3} \times 10^6 \\ &= 0.99 \times 10^4 \cong 1 \times 10^4\ \text{ppm} \end{aligned}$$

Approximation:

$$\begin{aligned} \text{conc. (ppm)} &= (0.1\ ml/10\ ml) \times 10^6 = 10^{-2} \times 10^6 \\ &= 1 \times 10^4 \\ \text{Relative error} &= ((1 \times 10^4)-(0.99 \times 10^4)/(0.99 \times 10^4)) \times 100 \\ &= ((0.01 \times 10^4)/(0.99 \times 10^4)) \times 100 \\ &= 0.0101 \times 100 = 1.01\% \end{aligned}$$

Solids in liquids

1 mg/ℓ expressed as g/mℓ is:

1 mg/ℓ = 0.001 g/1000 mℓ = 1 part/1,000,000 parts = 1 ppm

= $(1\ \text{part}/(1 \times 10^6))\ 10^6$ = 1 ppm

1 µg/mℓ expressed as mg/µl is:

1 µg/ℓ = 0.001 mg/1,000 µl = 1 part/1,000,000 parts = 1 ppm

= $(1\ \text{part}/(1 \times 10^6))\ 10^6$ = 1 ppm

Therefore: mg/ℓ = 1 ppm; µg/ml = 1 ppm; µg/ℓ = 1 ppb;

ng/ml = ppb; ng/µl = 1 ppm

latter restriction on sample injection is necessary (1) to minimize band spreading as the sample migrates through the column, and (2) to be certain all sample components begin the separation process at the top of the column as contrasted to being spread over several column volumes at the initial time t_o. More details about the various sample injection techniques outlined below may be found in discussions by Debbrecht (ref. 37), Schill (ref. 38), Cavagnol and Betker (ref. 39), Szonntagh (ref. 40) and Lebbe (ref. 41).

5.3.1 Syringe injection

The most common mode of liquid sample introduction is by direct syringe injection into the column carrier gas stream. For liquid samples the chromatographer uses a 5.0 or 10.0 µl syringe and normally injects 1.0-2.0 µl of sample. The injection of a 1.0 µl sample from a 10.0 µl syringe may introduce 2-5% error in the actual volume released. This error is caused by the additional sample volume held back by the needle (ca. 0.8 µl) and the time necessary to inject, depress the plunger and remove the syringe from the injection port. The common syringe for liquid samples has a glass barrel and a metal or polymer-coated metal as the plunger. Liquids sampled under high pressure should have metal barrels (e.g., brass or stainless steel). If the sample must be injected either cold or at a high temperature, the barrel may be enclosed in a jacket which may have either cooling or heating liquids circulating through it.

Gas samples may also be injected by syringe; however, a gas-tight syringe is necessary. The authors have found that gas-tight sampling syringes with valves such as those manufactures by Precision Sampling Corporation (Baton Rouge, LA, U.S.A.), are more convenient and result in better repeatability of sample volumes than gas-tight syringes which do not incorporate push-button valves at the bottom of the barrel. Proper aliquot removal with this gas syringe is ensured if the following steps are followed: (1) insert syringe (with syringe valve open) into the sample container, (2) pull plunger back to a barrel marking in excess of the desired sample volume for injection, (3) slide the plunger back and forth several times to minimize a sample aliquot fractionation, (4) set the plunger at the barrel marking in step 2 and close push-button valve at base of syringe barrel, (5) compress the plunger to required sample volume marking and open and close the valve to equalize (relieve excess) pressure in the syringe. This type of syringe is recommended when performing headspace analysis (see Section 4.2.2.1).

5.3.2 Gas sampling valves

There are numerous gas-sampling inlet systems available to the environmental scientist. Systems are available for sampling gases from liquids, from systems under high or low pressure, and corrosive samples. We will mention the various systems which have been used. The details of operation may be found in the references cited. A number of these sample systems have very little applicability to environmental samples, but their availability and application are useful information.

By-pass sample systems are of two types: (i) constant volume pipets and (ii) by-pass sample valve for syringe injection.

(i) Constant-volume pipets (refs. 42 and 43). These may be made of glass or stainless steel. The "pipet" is a detachable piece of tubing which is easily mounted in the system. The volume of these pipets can be from 0.1 ml (100 μl) to 10.0 ml. Depending on the volume necessary, the "pipet" may be made from capillary tubing, large I.D. tubing or a bulb. They are similar to gas sampling valves in that they must be flushed with sample prior to sample introduction. This restricts their use to large volume samples. Another disadvantage is the need for a pressure measurement and volume calibration.

(ii) By-pass sample valve for syringe injection. Syringe injection (Section 5.3.1) may be combined with a constant volume pipet (see above). The pressure in the by-pass loop is reduced to atmospheric conditions before injection of the sample. Small error is incurred if the syringe is leak-proof; the gas pressure in the loop is carefully controlled and the injection technique remains constant.

5.3.3 Automatic sample injection

The application of gas chromatography to process and/or quality control analyses requires dedicated instrument(s). A number of restrictions are imposed upon systems of this type. They should be able to be activated by remote controls, be able to operate overnight or over a weekend, must be leak-proof, and should not have valve systems easily contaminated. The valve systems for these sample systems work by electrical controls, electromagnetic controls or electromechanical controls. The type of valves used in these sample injection systems may be piston-type valves, diaphragm valves, rotating-plate valves or linear-sliding valves. Automatic sample-injection technique can be applied to liquid, solid, and gas samples. The reader is referred to the information provided by Shill (ref. 38) and Jeffery and Kipping (ref. 42) for additional information and details of the sampling systems.

5.3.4 Miscellaneous injection systems

Several injection systems which are for specific types of samples have been discussed in the literature. Since these do not have universal application to many sample types, they will be mentioned with little discussion. The environmental analyst having use of such systems should consult the original literature.

Glew and Young (ref. 44) designed a liquid nitrogen trap injection system for small amounts of condensable gases. A trap that is more practical (does not have to be mounted on the gas chromatograph) was used by Russell and Bednas (ref. 45). The former system requires mounting on the instrument as well as flushing the sample (once it has reached room temperature) from the U-tube sample holder onto the column, whereas the latter system involves injecting the sample through a septum with the needle located at the base of the sample chamber.

It is possible to generate the gas sample in the gas chromatograph. Any volatile gas, e.g., CO_2, H_2S, NH_3, which can be liberated from a sample by use of the proper reagent (acid or base) inside a small reactor connected to the gas chromatograph may be determined in this manner. Using this technique, the detector response may be calibrated by the same technique using a pure compound.

The components of interest may be removed from a stream of gas by adsorption on an inert material, e.g., activated charcoal or Tenax-GC and then released by heating in a desorption chamber attached to the gas chromatograph. Small concentrations may be determined by passing a large volume of sample through the adsorbent tube (see Chapter 3, Section 3.4.1).

The analysis of solid samples presents several options to the analytical chemist. He may inject the solid sample directly into the gas chromatograph, dissolve it in a suitable solvent and treat it as a liquid sample, deposit it on a piece of stainless steel or platinum gauze or encapsulate it in a piece of low-melting metal (e.g., wood's metal, m.p. 60.5°C) or glass (ref. 45A). Solid sample injection is not used to any great extent for environmental samples. Most solid samples are usually extracted with an appropriate solvent and then treated as a liquid sample. Employing this technique places restrictions on the analyst. The solvent used must (1) not react with any of the components of the sample, (2) not leave any solid residue remaining unless one is sure that the components of interest have been dissolved, and (3) not elute with any of the components of interest. The technique of deposition of the solid sample on a piece of stainless steel or platinum gauze has the advantages that (1) solvent peaks will not interfere with the components of interest, (2) dilute solid samples may be concentrated on the gauze, and (3) non-volatile components usually will not enter the column and cause contamination. Encapsulating solid samples in either a low melting metal or glass is used when the sample volatile materials are a part of the analysis scheme.

5.4 COLUMNS AND COLUMN SELECTION FOR SEPARATION AND ANALYSIS

The variety of columns available to the environmental chemist is confusing, at the least. Analytical chemists who have been working in the area of separations have their special columns for solving most environmental problems but the novice becomes bewildered when selecting a column for a particular type sample. If the sample he has is governed by some regulatory agency's procedure, the column and conditions are predetermined. The acceptance of procedures published by governmental agencies can result in a laboratory stocked with numerous types and lengths of columns.

The wise analytical chemist will have a stock of "dependable" columns to meet most problems that may arise. A "dependable" inventory means different things to individual chemists but one such "inventory" of columns would include: (1) a molecular sieve column (5A or 13X) for separation of gases such as H_2, O_2, CO and CH_4, (2) a porous polymer column (Porapak or Chromosorb Century series) for the separation of gases and low molecular weight components, (3) Apiezon L column for high boiling aliphatic and aromatic hydrocarbons and similar non-polar type sample components, and (4) a Carbowax 20M column for phenols, pesticides and halogenated compounds.

The inexperienced chromatographer is not completely at a disadvantage. First, if he needs to separate a mixture he should know qualitatively the make-up of the sample. Starting with this information and following the general rule that "likes dissolve likes" he can proceed to simplify his problem.

Rohrschneider (refs. 46 and 47) initially chose three compounds (probes) to represent various solute-solvent interactions for the process of solution: ethanol for donor-type interactions, nitromethane for dipole-type interactions and pyridine for acceptor-type interactions. When his polarity scale was broadened to characterize stationary liquid phases, two more probes were added: benzene and 2-butanone. These two compounds extended the interaction type processes to include electron density and hydrogen bond acceptor, respectively.

McReynolds (ref. 48) added five more probes and changed three of the original Rohrschneider probes, i.e., 2-pentanone for 2-butanone, 1-nitropropane for nitromethane and 1-butanol for ethanol. The basis of the McReynolds constants is that these ten probes (compounds) are compared on a non-polar stationary phase (squalane) and the stationary phase in question (considered more polar than squalane). The difference in the adjusted retention time or adjusted retention volume of the probe on the two columns expressed as retention index, I, is used to tabulate the ΔI(McReynolds constant) values. The basis for the retention index system (see Section 5.6.1.2) is a homologous series of even-numbered carbon atom n-alkanes. The retention index of each n-alkane is 100 times the number of carbon atoms in the molecule, i.e., hexane = 600, octane = 800, etc.

These McReynolds constants have been tabulated for all the common commercially available liquid phases. Most chromatographic suppliers furnish the McReynolds constants for all liquid phases they sell. To use these constants one simply matches the type compounds to be separated with the appropriate probe compounds and selects the liquid phase with the highest McReynolds constant. Table 5.5 matches the ten-probe compounds with sample compounds which would be expected to behave similarly.

Liquid phases which have equivalent McReynolds constants for all ten probes compounds are similar type phases: e.g., SE-30, OV-1, OV-101 and DC-200 (all methyl silicone) or OV-17 and SP-2250 (both methyl/phenyl silicones). Most books on gas chromatography have a section discussing column selection and the use of the McReynolds constants; an excellent book on this topic is the one by Supina (ref. 49). The tables in the appendix of Supina's book list all the liquid phases in numberical order according to each probe compound and also alphabetically by liquid phase. These tables give this book an advantage over other listings of the McReynolds constants.

Glass open-tubular capillary columns (OTC) have become a standard piece of equipment in many laboratories and are no longer considered to be a "novel" column. Grob and Grob (ref. 50) compared two chromatograms of the same lake water extract on a packed and a wall-coated open tubular (WCOT) glass column. Although the capillary

TABLE 5.5

McReynolds probe compounds compared to sample compounds

McReynold compound	Sample compounds which should behave similarly
Benzene	Aromatics and olefins
1-Butanol	Alcohols and phenols (weak acids)
2-Pentanone	Aldehydes, ketones and esters
1-Nitropropane	Nitro and nitrile compounds
Pyridine	Bases and aromatic heterocycles
2-Methyl-2-pentanol	Alcohols and branched chain compounds
1-Iodobutane	Halogen compounds
2-Octyne	Acetylenes and possibly olefins
1,4-Dioxane	Ethers and bases
cis-Hydrindane	Non-polar steroids, terpenes and naphthenic compounds.

column was 12 times as long as the packed column, there was only 1% as much stationary phase on the walls of the WCOT column. As a result, smaller sample sizes had to be used with the capillary column as with the packed column (0.010 µl for the capillary column to 2.0 µl for the 0.125 in. O.D. packed column). Four times as many component peaks (very narrow and symmetrical) appeared from the WCOT column as with the packed column, meaning higher separation efficiency. Baseline stability of the capillary separation is an indication of nearly complete sample resolution, optimum experimental conditions and thermal stability. Thin-film, narrow-bore columns (0.25 mm I.D.) are more efficient than large-bore (0.5 mm I.D.) columns. However, the efficiency of narrow-bore columns decreases rapidly (i.e., number of theoretical plates per meter decreases) as film thickness increases (> 1.5 µm). The large-bore capillary columns (0.5 mm I.D.) can accommodate stationary phase films up to 2.0-2.5 µm and thus are able to be used with direct injection techniques, larger sample sizes, and a larger sample concentration range. These added advantages do bring disadvantages: lower column efficiency (ca. 10^3 plates per meter) and longer analysis times. The outstanding advantage of capillary columns versus packed columns is that the choice of stationary phases becomes less of a problem because their efficiency is so high. This translates into the fact that a smaller number of phases will be necessary, i.e., a nonpolar phase (e.g., SE-30 or OV-101), a polar phase (e.g., Carbowax 20M), and an intermediate polar phase (e.g., OV-17).

This does not imply that packed columns will disappear from the laboratory. They will always have a place for the separation and analysis of samples containing 10 components, low molecular-weight gases, preparative work and separations coupled with detectors (e.g., infrared or nuclear magnetic resonance) for low sensitivity.

The reader is referred to a number of excellent books concerning capillary columns in order of their publication: Open Tubular Columns in Gas Chromatography by

Ettre (ref. 51), Gaschromatographie, by Schomberg (ref. 52), High Resolution Gas Chromatography by Freeman (ref. 53) and Gas Chromatography with Glass Capillary Columns by Jennings (ref. 54).

An important concern in capillary chromatography is quantitation. Because the most common mode of sample introduction onto capillary columns is with valves in which a syringe-injected sample is dynamically split in a heated tube between the column and a vent, discrimination of sample components can occur. This problem is usually seen with wide boiling-point range samples and/or polar compounds. Increasing the inlet temperature may help in some cases, but it is not a cure-all for this problem. The cold on-column injection into an unheated injection port has been demonstrated to be indiscriminate as far as the sample components are concerned (ref. 54A). A specially designed inlet for cold on-column injection is needed. (This is commercially available from Hewlett Packard Co., Avondale, PA and Carlo Erba Strumentazione (refs. 54B and 54C)). Unfortunately, this cold on-column injection method cannot be automated. (See Appendix II.)

5.5 DETECTION OF SAMPLE COMPONENTS

A number of detectors are available to the environmental analytical chemist for use in gas chromatography (refs. 55-57). The two most widely used detectors are the flame ionization detector (FID) and the electron capture detector (ECD). The flame ionization detector destroys the sample during its measuring process; however, by use of gas stream splitter, a predetermined percentage or ratio of the column effluent can be diverted away from the detector for further study and/or identification. Interfacing the gas chromatograph with a mass spectrometer (ref. 58) provides a means of both identification and quantification.

To obtain good gas chromatography-mass spectrometry (GC-MS) operation, it is necessary to optimize the gas chromatographic method, the mass spectrometric techniques, and the interface techniques. As a first estimation, a differentially pumped mass spectrometer can receive 1 cm^3 atm/min (1.3×10^{-2} liter Torr/sec) of the carrier effluent gas. The mass spectrometer can accommodate the carrier effluent from either packed or open-tubular columns. To achieve good qualitative analysis with a gas chromatographic-mass spectrometric system, pure chemicals are necessary. The best gas chromatographic-mass spectrometric data come from chromatography systems which have high column efficiency prior to the mass spectrometric analysis. This high column efficiency is not always necessary because unresolved peaks can, in many cases, be identified by GCMS. The tailing of peaks from polar compounds can be very troublesome using this technique, especially if the compounds being eluted are chemically similar (alcohols and aldehydes are the most troublesome). This problem may be minimized by adding surfactants to the liquid stationary phase. Ionic surfactants probably react chemically with the adsorption sites of the column, whereas non-ionic surfactants probably function more as wetting agents. Some common surfactants used are: Igepal CO880, Alkaterge T, and Span 20. The carrier gas used in GC-MS must (1) be chemically inert, (2) not interfere with the mass spectrometric pattern, (3) enable enrichment of the sample components in the gas stream, and (4) not interfere with the total ion detection. The most commonly used carriers in GC-MS are helium, hydrogen, and nitrogen. Helium is the best. Hydrogen interferes with total ion detection and nitrogen does not always enrich the sample components in the carrier stream, interferes with the total ion detection, and interferes with mass spectral patterns in the low mass range. Additional qualitative information may be obtained when specific chromatographic detectors are used in combination with the mass spectrometer. For example, a flame photometric detector for compounds containing phorphorus and/or sulfur, the coulometric detector for compounds containing sulfur, nitrogen and/or halogens, the thermionic detector for compounds containing phosphorous, halogen and/or nitrogen, and the electron capture detector for compounds containing halogen and/or sulfur as well as functional groups, i.e., conjugated carbonyls, di- and trisulfides, and nitriles.

Gas chromatographic detectors may be placed in two categories: (1) detector response is concentration-dependent, or (2) mass-flow-rate dependent. The sensitivity is the ability of the detector to respond to compounds entering its environment. The sensitivity of concentration-dependent detectors is expressed as:

$$\text{Sensitivity} = (\text{peak area} \times \text{flow-rate})/\text{sample weight} \tag{5.17}$$

$$= (\text{mV} \times \text{cm}^3/\text{min})/\text{mg} \tag{5.18}$$

The sensitivity of mass-flow-rate dependent detectors is expressed as:

$$\text{Sensitivity} = \text{peak area}/\text{sample weight} \tag{5.19}$$

$$= \text{A}/\text{g} \tag{5.20}$$

The lower limit of detection (LLD) is the smallest amount of sample which will cause a measurable signal (i.e., twice the noise) over the noise signal. This is also known as the minimum detectable limit (MDL) or minimum detectable quantity (MDQ). Detector specificity is the ratio of the detector response of a contaminant (interfering substance) to that of the desired component. Detector linearity is the range over which the detector maintains constant sensitivity to increasing concentration of a specific component, i.e., it is the range over which the response is directly proportional to the mass flow-rate or the concentration of a component. If a larger quantity passes through the detector the output is no longer proportional and the detector is "saturated." The component peak in the non-linear range will be rounded at the apex. Detectors are also described by their response time (time necessary for the detector signal to reach 63% of its true value). A recorder which has a response time slower than that of the detector limits the detector response time.

Table 5.6 summarizes some of the more widely used detectors. The analytical chemist must choose the detector on the basis of the type of compound to be quantified and its concentration range in the samples.

TABLE 5.6

Widely used gas chromatographic detectors

Type	Components detectable	Linearity	Response category	Minimum detectable limits	Sample compound
Flame ionization (FID)	Organic compounds	5×10^6- 7×10^7	MFR*	10^{-12} g(C)/sec.	Propane
Electron capture (ECD)	Halogenated, oxygenated compounds	10^1-10^2	C**	10^{-16} moles/cm^3	Lindane
Flame photometric (FPD)	Phosphorous and sulfur compounds	5×10^2(S) 1×10^3(P) log/log scale	MFR*	10^{-10} g(S)/sec 2×10^{-12}g (P)/sec	Thiophene Tributyl-phosphate
Thermionic/alkali flame (TID)/(AFID)	Phorphorous and nitrogen containing compounds	10^3	MFR*	5×10^{-14} g (N)/sec 5×10^{-15} g (P)/sec	Azobenzene Tributyl-phosphate
Thermal conductivity (TCD)	Organic, inorganic, compounds	10^4	C**	5×10^{-10} g/ml	Propane

*Mass flow rate dependent, (mv/mg/sec) ** Concentration dependent, (mv/mg/ml)

Permanent gases may be detected in the ppb range with the use of a helium detector. A beta-source (250 mC) excites the helium atoms which in turn excite molecules having an ionization potential lower than that of helium. Representative ionization potentials are shown in Table 5.7. If the gases to be determined are in the ppb range, a 400 V electric field will yield measurable ion currents of about 4 x 10^8 A. The helium detector requires highly purified helium for the carrier gas and has a sensitivity higher than the thermal conductivity cell (TCD) or the flame ionization detector (FID).

TABLE 5.7

Ionization potentials (eV) of selected gases

H_2S	10.4	CH_4	14.5
O_2	12.5	N_2	15.5
SO_2	13.1	H_2	15.6
CO	14.1	Ar	15.7
CO_2	14.4	Ne	21.5
He	24.5		

5.6 QUALITATIVE AND QUANTITATIVE INFORMATION

The realm of qualitative and quantitative analysis in GC is much broader than the "user" of chromatography realizes. To the analytical chemist these two divisions of analysis are very specific categories. In the general sense of the word, qualitative denotes the kinds of components present, whereas quantitative conveys the amount of each component that is present. Obtaining quantitative data from a chromatographic system is a straightforward process once the chromatographer has established accurate qualitative data.

There is no error-proof way of obtaining reliable qualitative information solely from the chromatogram. There are a number of very basic reasons for this. (1) The number of peaks present from an unknown sample is no guarantee of the number of components in the sample. Several components could have the identical retention time (or volume) under the conditions that the separation was obtained. (2) The assumption that one makes is that the entire sample was transported through the column. Part of the sample may be deposited at the top of the column and never moved down the column. (3) Improper conditions (i.e., temperature, pressure, column stationary phase, etc.) may have been used and either pyrolysis of the sample components or reaction of the sample components with the column resulted. What this means to the analytical chemist is that each peak must correspond to a single sample component. The authors wish to emphasize this point because we have seen many persons come to the wrong conclusion using the chromatogram as their sole source of information. If the chromatographer succeeded in separating each and every sample component, there is no absolute method of establishing exactly the correspondence between peaks and sample components from retention data alone.

5.6.1 Qualitative analysis by gas chromatography

Qualitative data in gas chromatography are obtained from two general categories of information: (1) information obtained from retention data and (2) information obtained by employing auxiliary techniques. Retention data are a function of the partition coefficient (a thermodynamic function); however, they may not be too informative unless the component make-up of a sample is known or the components are all members of an homologous series. The use of retention data to identify components of a complete unknown sample leaves much to be desired.

5.6.1.1 Retention data. Retention data in terms of time, i.e., retention time (t_R) and adjusted retention time (t'_R) are too dependent upon column parameters. It is better to use retention volume, V_R.

$$V_R = t_R F_c \tag{5.21}$$

adjusted retention volume, V'_R

$$V'_R = (t_R - t_M)F_c \tag{5.22}$$

net retention volumn, V_N

$$V_N = jV'_R \tag{5.23}$$

or specific retention volume, V_g

$$V_g = (V_N/Tw_L)273 = ((jV'_R)/(Tw_L))273 \tag{5.24}$$

where F_c = volumetric flow rate of mobile phase at column outlet in ml/min.,

t_M = retention time of an unretained peak,

j = pressure gradient correction factor = $3/2\ ((p_i/p_o)^2-1)/((p_i/p_o)^3-1)$; p_i and p_o are the mobile phase inlet and outlet pressures;

T = column temperature (°K), and

w_L = weight of stationary phase in column,

because these parameters are less dependent upon column conditions. The more reliable term for qualitative analysis is relative rentention volume, $r_{a/b}$.

$$r_{a/b} = (V'_R)_a/(V'_R)_b = (V_N)_a/(V_N)_b = (V_g)_a/(V_g)_b \tag{5.25}$$

because one is comparing the retention volume of a component to the retention volume of a standard substance under the same column conditions. The choice of the standard is dependent upon (1) its availability and (2) its retention compared to other components in the sample (should be half-way in the range of retention volumes being compared).

Additional qualitative information may be obtained if one separates the sample components on two columns of different polarity and plots the log adjusted retention volume (V'_R) versus the reciprocal of absolute column temperature. The slopes of the data lines from each column usually are different and even the order of elution of some components change.

5.6.1.2 Retention index. The retention index of Kovats (refs. 59-62) compares retention data to a series of standards, rather than one compound standard. The series of standards was the even-carbon-atom n-alkanes and their index, by definition, was 100 times the number of carbon atoms in the molecule, i.e., ethane was 200; hexane was 600, etc. This index system is described mathematically as

$$I = 100N + 100(\log V'_R(A) - \log V'_R(N))/(\log V'_R(n) - \log V'_R(N)) \quad (5.26)$$

where A = unknown component;
N and n = the smaller and larger n-alkanes which bracket the unknown component A.

The n-alkanes which are chosen should give peaks completely separated from the sample component. Hupe (ref. 63) has constructed a nomogram from which one may calculate I values without doing the lengthy calculations.

Polar compounds display greater temperature dependence than do non polar compounds on the Kovats' index scale. When reporting temperature dependent data it is useful to give the temperature coefficient of the I value in reference to a 10° C temperature change plus the temperature range covered, e.g.,

$$\delta I/10°C = 4.7 \quad (5.27)$$

Some general guidelines for using this index from data collected on a single stationary phase are: (1) as the number of carbon atoms increases in an homologous series, the retention index increases by 100 units, (2) the same functional groups on compounds having similar structures increase the retention index to the same extent, (3) a simple relationship exists between boiling point and retention index on a nonpolar phase (thus index systems can be used to predict boiling points of unknown components) and the difference between the retention indices of isomeric compounds can be approximated by the difference in boiling points ($\delta I = 5\delta$ b.p.).

Guiochon (ref. 64) extended Kovats' work on the index system to molecular structure and concluded that (1) I values of nonpolar compounds are usually constant using any stationary phase, (2) I values for some components on different polarity phases are characteristic of molecular structure, (3) I value of any compound is the same on nonpolar stationary phases, and (4) the presence of unsaturation sites, functional groups and rings in a compound add incremental values to the basic retention index.

The Kovats retention index is more accurate than other retention data parameters, provided the support material causes no adsorption effects with polar compounds, the column temperature and the mobile phase flow rate are controlled, and the sample size is in the linear portion of the partition isotherm. Reproducibility of retention indices are very good for nonpolar compounds and within 2% for polar compounds.

5.6.1.3 Auxiliary techniques. The use of retention data does not give the analytical chemist unequivocal identification of sample components so other techniques must be employed. This additional information may be obtained from modification of the chromatographic system per se or by employing measurements of detection other than the usual chromatographic detectors.

(i) Chromatographic measurements may be obtained by (1) using a specific detector or a combination of detectors or (2) using other chromatographic systems. As pointed out in Section 5.5, various detectors respond to certain types of compounds or functional groups, e.g., electron capture detectors for compounds containing halogens, the flame photometric detector for phosphorus- and sulfur-containing molecules, and radioactive detectors for the detection of labeled compounds (ref. 65). Additional information may be obtained by doing a preliminary separation using liquid chromatography (Chapter 6 and ref. 66), thin-layer chromatography (ref. 67) or paper chromatography (ref. 68). Golovkin et al. (ref. 69) combined retention indices and selective detectors for the qualitative identification of halogen, phosphorus and sulfur containing pesticides.

(ii) Qualitative information may be obtained by combining other detection techniques than the usual chromatographic detectors. The combination of GC and MS is a widely used system for qualitatively identifying components in environmental samples (ref. 70). Dioxin in tissue has been determined by the combination of capillary GC and atmospheric pressure negative chemical ionization (ref. 71); mirex pollution in water has been assayed by using capillary GC-MS (ref. 72), and the light fractions in tar have been identified using GC-MS (ref. 73). Infrared spectro-chemical detection was used by Baranski et al. (ref. 74) to study the catalytic properties of NaHY zeolites, whereas cresols in work area air were detected by Vistocco and Beggio (ref. 75). Sulfur compounds in flue gases were identified and determined by non-flame, source-induced sulfur fluorescent detectors (ref. 76) and Chow and Karmen (ref. 77) determined primary amines by on-line fluorescence measurements. Derivatives were made of amino acids (oxazolidinones) and measured by flame ionization detection (ref. 78). Pannatier and Testa (ref. 79) have written a review article about the use of derivatives to aid in the identification of components in unknown mixtures. Another ancillary technique which may be combined with GC is that of pyrolysis. Kessel'man and Ogloblina (ref. 80) did a preliminary separation of organic compounds with thin-layer chromatography and then pyrolyzed the spots followed by GC. Eight Bacillus microorganisms in cells were identified by pyrolysis GC by Chou (ref. 81).

5.6.2 Quantitative analysis by gas chromatography

Quantitative GC is concerned with how much of a component is present in a sample. For a more detailed coverage of the various techniques of obtaining quantitative data, the reader is referred to the excellent chapter by Debbrecht (ref. 82) and the book by Novak (ref. 83).

All the commonly used detectors in GC (TCD, FID, ECD, FPD, etc.) produce a differential chromatogram, i.e., if conditions are ideal and the column effluent flows through the cell uniformly, one obtains peaks which are gaussian in shape. These may be mathematically described by eqn. 5.28:

$$n = 0.5m\ (\pi Dt)^{-1/2}\ e^{-l^2/4Dt} \qquad (5.28)$$

where n = concentration of gas molecules;

m = total quantity of diffusing material;

π = constant = 3.1416;

D = diffusion coefficient;

t = time; and

l = distance molecules diffuse.

The size of the peak is proportional to the amount of component in the sample. If this is true, the peak area and/or the peak height will be proportional to the amount of the component causing the peak. Peak height is proportional to the amount of a component contributing to the peak if there is no change in the system which will cause a change in the peak width when comparing a sample to a standard. Four parameters may influence the peak width: (1) temperature, (2) flow rate of the mobile phase, (3) sample injection, and (4) column overload. To maintain an accuracy of 1% in the analysis the temperature must be controlled to within $\pm$0.3°C, preferably $\pm$0.1°C and the flow rate controlled to within $\pm$0.1%. Poor sample injection will have an effect on peak width; peaks which emerge quickly (i.e., 1-2 times the hold-up volume) are affected the most. Peaks which are at least ten times the hold-up volume are affected the least. Component adsorption on the column affects peak width the most when the concentration is low. When too large a sample is injected, the stationary phase becomes saturated, which broadens the peak and consequently reduces peak height.

The peak height or area may be obtained by manual methods or by use of an integrator. Fig. 5.1 illustrates the constructions necessary for manual measurements. A line AB is drawn along the base of the peak so that it connects both sides of the peak to the baseline. A perpendicular to line AB is drawn from the peak maximum and labeled CD. At a point halfway up the perpendicular CD, another line EF is drawn parallel to the base AB. This last line is known as the peak width at half-height. The product of CD and EF is the area of the peak.

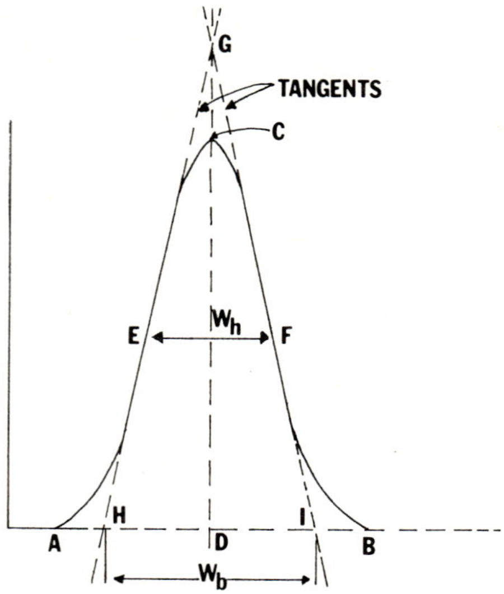

Fig. 5.1 Lines and points of construction for peak size measurements.

Integrators receive the detector response signal and integrate it directly on a time basis (provided the number of counts per unit time are proportional to detector response) which gives a simple relationship between the total number of peak counts and the peak area. This can be shown mathematically by eqn. 5.29:

$$n_i = KA_1B^{-1}(db/dt)^{-1} \qquad (5.29)$$

where n_i = total number of integrator counts;
K = proportionality constant;
B = recorder constant
db/dt = recorder chart speed; and
A_1 = peak area.

Returning to Fig. 5.1 one sees that there are three measurements which are proportional to the amount of a component from the chromatographic peak:

(1) the peak height, CD;

(2) the peak area measured by the product of peak height CD and the peak width at half-height EF, and

(3) the peak area measured by triangulation, i.e., the area of the triangle GHI, which is GD times one-half the base HI.

Once the chromatogram has been interpreted by use of peak height or peak area, it is necessary to use some standardization technique to relate one or both of these measurements to standards. There are several techniques available for standardization. We shall not discuss each one and/or the variants of the techniques. Alternatively, we shall present the more widely used methods for environmental analysis. More detail may be obtained from refs. 82 and 83 or from a three-part article in American Laboratory by I. G. Young (ref. 84).

(i) Internal standardization. A standard substance is chosen that (1) elutes from the column separated from all components of the sample, (2) elutes close to the desired component under study, (3) has a similar functional group(s) to that of the desired component, (4) is stable and unreactive with sample components, or (5) is non-volatile enough that standard solutions may be made and stored for long periods of time. Several standard solutions containing known weights of the chosen standard and desired component to be studied should be prepared. The weight range chosen must not overload the column or the detector. A plot of weight ratio versus area ratio for these standard solutions should be linear. To determine the amount of the desired component in the sample, a known weight of the standard is added. The response factor, F, of the desired component is calculated from any of the standard solutions prepared and the following equation:

$$F = \frac{\text{weight of desired component/weight of standard}}{\text{peak area of desired component/peak area of standard}} \quad (5.30)$$

The weight of the desired component, W_c, in the unknown can then be calculated using eqn. 5.31:

$$W_c = (A_c \times W_i)/(A_i \times F) \quad (5.31)$$

where W_i and A_i are the weight and peak area of internal standard in the sample, A_c is the peak area of the desired component, and F is the response factor of desired component (calculated from eqn. 5.30).

(ii) External standardization. A calibration curve is prepared (using identical conditions as the sample) for each substance in the sample which is to be quantitated. This curve may be (1) peak area versus concentration or (2) peak height versus concentration. When the sample is run, the peak height or peak area of the substances to be quantitated is compared to the calibration curves.

(iii) Internal normalization. To use internal normalization, each and every sample component must produce its own individual peak. One of the sample components may be used as the reference, or another component which is separated from all sample components may be used. A standard sample is made up containing known weights of each component. The ratio of peak area (or peak height) of each substance to actual weight percent in standard sample is calculated for each component. The reference component is assigned a response factor, F, of unity or 100 (ref. 85), and the response factor of the other components in the standard are calculated by dividing the ratio of each component by the ratio of the reference component. Table 5.8 illustrates data for a standard sample and Table 5.9 illustrates data for an unknown sample using the response factors from Table 5.8

TABLE 5.8

Internal normalization of standard sample

Component	Weight used(mg)	Weight %	Peak area	Area %	Area wt.%	Response factor,F
1	354.3	21.67	3901	22.22	180.0	1.000
2	296.7	18.15	3654	20.81	201.3	1.118
3	542.3	33.17	5236	29.82	157.8	0.877
4	441.5	27.01	4768	27.15	176.5	0.981
Total	1634.8	100.00	17559	100.00		

TABLE 5.9

Calculation of weight percent of each component in unknown sample

Component	Peak area	Area %	Area F	Weight %
1	4265	25.45	4265	24.71
2	3231	19.28	2890	16.74
3	5496	32.80	6267	36.31
4	3765	22.47	3838	22.24
Total	16757	100.00	17260	100.00

(iv) Standard addition. Internal standardization (see i above) and standard addition are very similar techniques; however, the standard addition technique requires no new component to be added to the sample. In fact, the component to be determined is the standard which is added. Standard addition is also known as "spiking" a sample. Two injections of the sample are necessary to use the quantitative technique. There are several variations of the standard addition technique; (1) measurement of the sample component of interest, (2) measurement of another reference substance in the sample, and (3) measurement of an added reference substance to the sample. For this discussion we are only concerned with the system where a sample component is used as the standard substance, i.e., variation 1.

The first step is to inject a known amount (i.e., volume, v) of the chromatographic unknown. The component of interest, c, will have a peak area A_c. If the concentration of component c, is expressed in molarity or in weight, the technique may be used. The sample is then "spiked" with a known amount of the component to be determined. Now the component c will have a peak area, A_{cs} (sub s means standard added). This increased peak area is due to the amount of "spike" added. Knowing the peak areas and the amount of standard component c added, one is able to calculate the amount of c in the original sample.

Assume 1.0 μl of an unknown sample (total volume = 1.0 ml) produced a peak of area equal to 40.0 mm^2. To the sample one add 10.0 μl of the pure component having a density of 0.900 g/ml.

$$(0.900\ g/ml) \times 0.010\ ml = 9 \times 10^{-3}\ g = 9\ mg \qquad (5.32)$$

The original volume of sample was 1000 μl (1.0 ml), less the 1.0 μl injected give 999 μl of original sample. If the volumes of sample and added standard are additive, the final "spiked" solution has a volume of 1009 μl, i.e., 9×10^{-3} g/1009 μl or 9.00 mg/1009 μl of "spike". 1.0 μl sample injection of the "spiked" solution produces a peak of 150 mm^2. This means the increase of 110 mm^2 is due to 8.92×10^{-3} mg of component c. Thus, the original sample injected corresponded to 3.24×10^{-3} mg of component c (40 mm^2 × (8.92×10^{-3} mg)/110 mm^2). So the original sample (1.0 μl) contained:

$$(3.24 \times 10^{3}\ mg/\mu l)(1000\ \mu l) = 3.24\ mg\ \text{component c} \qquad (5.33)$$

5.7 ANALYSIS OF AIR AND WATER CONTAMINANTS

In this section, we present a number of the more recent methods for the analysis of both air and water samples for many of the commonly encountered pollutants. The listing is by no means complete; however, it is representative. The methods are in two categories: (1) air samples, Table 5.10 and (2) water samples, Table 5.11. No detailed evaluations are presented for any of the methods. The main justification of such a section is to give the reader a point of origin at which he may find a discussion of one or more of his problems. In addition, we have not given any references prior to 1978, unless they represent an existing present day state-of-the-art methodology. The reader desiring more information about gas chromatography methodology for air and water samples may consult several very good sources: (1) Chemical Abstracts published by the American Chemical Society, Columbus, OH, USA, (2) Gas and Liquid Chromatography Abstracts published by the Chromatography Discussion Group of Great Britain, Trent Polytechnic, Nottingham, and (3) The Preston Technical Abstracts published by the Preston Technical Abstracts Company of Niles, IL, USA. Two other excellent sources of information are the Application Reviews of Analytical Chemistry (ref. 86) which are published in the odd-numbered years and the Fundamental Reviews of Analytical Chemistry (ref. 87) which are published in the even-numbered years. Both are published by the American

Chemical Society, Washington, DC, USA. The Fundamental Reviews cover the topic of gas chromatography whereas the Application Reviews cover air pollution, water pollution, and pesticide residues.

TABLE 5.10

AIR SAMPLES

Compound(s) type	Column(s)	Detector(s)	Sample type	Detection limit(s)	Reference
Tetra-,Hexa-, Hepta-,Octa-chlorodibenzo-p-dioxin isomers	2 mm x 210 cm glass. 0.6% OV-17 +0.4% Poly S-179. 80-100 mesh Permabond methyl silicone.	Low-resolution MS	Fly Ash, Activated Sludge, Particulate Matter	Low ppt	88
N-nitrosamines	Basic Al_2O_3	Thermal energy analysis	Cigarette smoke	0.1 ng	89
Organophosphorus Pesticides	SE-30 Carbowax 20M	ECD, TID	Air	1-5 ng	90
Polycyclic aromatic hydrocarbons and naphthalenes	--	FID	Tobacco smoke	--	91
1,2-Dibromoethane	1.5% OV-17+ 1.95% OV-210	^{3}H ECD	Ambient air	3 pg	92
1,2-Dibromo-3-chloropropane	1.5% OV-17+ 1.95% OV-210	^{63}Ni ECD	Ambient air	0.02 ppb	93
Nitrogen containing compounds	--	MS	Cigarette smoke	--	94
Nitric oxide	--	ECD	Air	0.01 ppm	95
Waste gases from tetrachloroethane production. Cl, HCl, C_2H_2, CO_2	6 mm x 3 m Silokhrom S-120 = (1m) + Porokhrom 3 (2m) coated with di-butylphthalate	TCD	Waste gas	Rel. Error < 1%	96
s-Triazine herbicides	Carbowax 20M WCOT	FID, AFID	Environmental samples	5-10ng(1:90) 50-70pg(1:20)	97
Vinyl chloride, ethyl chloride, Tetrachloroethylene	Carbowax 1500 on Carbopack A	FID, Microwave plasma	Air	< 2 ppm < 1 ppm ~ 1 ppm	98
Pesticides	3% OV-17 on Chromosorb W-AW	ECD	Air	0.04-8.0 $\mu g/m^3$	99
Hydrogen cyanide	Porapak Q	AFID	Air in fire environment	~ 5 pg	100

TABLE 5.10 (Cont.)

Compound(s) type	Column(s)	Detector(s)	Sample type	Detection limit(s)	Reference
$CHClF_2$	Chromosorb 102	ECD	Ambient air	9 ppt	101
Benzene	10% oxydipropionitrile on porolith	FID	Air	$0.05 mg/m^3$	102
CO_2	Porapak R	TCD	HCl	~ 5 ppm	103
Styrene	10% FFAP on chromosorb W	FID	Air	0.33 µg/ml	104
Smoke condensate	6% OV-17 Chromosorb W	FID	Cigarettes	--	105
HCN	4mm x 3m Porapak Q	FID	Exhaust gas	< 0.1 ppm	106
Pentachlorophenol, tetrachlorophenol	Silicone DCFS-1265. Chromosorb WAW - DMCS	ECD	Air	--	107
Polycyclic aromatic hydrocarbons	SKT-30, Silochrom S-80	FID	Waste gases	1.1×10^{-6} mg	108
Acetone	15% Carbowax 1500, Chromaton N - AW	FID	Air	$2 mg/m^3$	109
Beryllium	3% SE-30, Diatomite CO	ECD	Air particulates	$<0.0001 \mu g/m^3$	110
SO_2	3.2mm x 2m Polyphenyl ether + H_3PO_4 on Chromosorb T.	FPD	Air	~ 2 ppb	111
Butyraldehyde	20% polyethylene glycol adipate on Chromatone N-AW-HMDS	FID	Air	$0.011 mg/m^3$	112
Bromoethane, chloroethane, ethanol, butanol, chlorobenzene	4mm x 1 m FSb on INZ 1200	FID	Air	0.01 µg	113
C_2H_5Br, $C_2H_4CL_2$ CH_3Cl	DC-550 + OV-17 + SE-30 on chromocorb WAW - DMCS	FID	Air	--	114
Chlorinated dioxins	--	MS	Fly ash	230 pg	115
Trimethylamine	--	FID	Air	0.2 ppb	116

TABLE 5.10 (Cont.)

Compound(s) type	Column(s)	Detector(s)	Sample type	Detection limit(s)	Reference
Methane, acetylene	3mm x 6m, 16% β,β-oxydipropionitrile	FID	Calcium carbide reaction gas	0.000294%(v/v)	117
Methyl chloride	2m Porapak Q	FID	Ambient air	11.9 ppb	118
Tetraethyllead cpds.	OV-101 on Gas-Chrom Q	Atomic absorption	Air	0.2ng/m^3	119
Aldehydes	1% PEG-HT on Chromosorb WAW - DMCS	FID	Air	0.3 ppb	120
HCN	--	TID	Auto exhaust	50 ppb	121

TABLE 5.11

WATER SAMPLES

Compound(s) type	Column(s)	Detector(s)	Sample type	Detection limit(s)	Reference
Twelve chlorobenzenes	1mm x 30m Carbowax-20M. 1mm x 30m SP-2100.	^{63}Ni ECD	Water	0.01ng/ℓ to 500ng/ℓ	122
Arenes, vinyl chloride, haloorganic-volatiles	3m Chromosorb 102. 2m, 20% SE-30 on Gas Chrom Q. 2m Porapak T.	ECD FID, PID	Water	1μg/ℓ	123
Ketones	4m x 3mm glass column. 5% Carbowax 1500 on Chromaton N-AW-HMDS	FID	Waste water	0.2mg/L	124
Volatiles from urea-formaldehyde glue plant	2.2% Ethofat 60/25 on Chromosorb 101.	FID	Waste water	1 - 5 ppm	125
Chlorophenols	--	ECD	River water Sediments	0.01-4.0ng/g	126
Aroclor 1016 & 1242	12 different phases	ECD, MS	Real samples	--	127
Nitrobenzene	5% OV-225 on Gas-Chrom Q. 2% Silicone DC-QF-1 on Gas-Chrom Q.	ECD	Water, Sediments Fish	0.01-1.0 μg/ℓ	128
Insecticides	6% QF-1 10% DC-200	ECD, FPD	River Water	<0.067μg/ℓ <0.5μg/ℓ	129

TABLE 5.11 (Cont.)

Compound(s) type	Column(s)	Detector(s)	Sample type	Detection limit(s)	Reference
CCl_4, $CHCl_3$	Silar 10-C, Gas-Chrom Q	ECD	Water	100 ppt	130
N-methylpyrrolidone	15% PEG-6000 + 4% KOH. Celite 545	FID	Waste water	0.005g/ℓ	131
Organochlorine insecticides, polychlorinated biphenyls	OV-1 + QF-1 on Chromosorb W. Apiezon L on Chromosorb G	ECD	Water	3-15mg/ℓ 106ng/ℓ	132
Bayer 73	OV-3, Gas Chrom Q	^{63}Ni-ECD	Mud water	--	133
Chlorinated pesticides	OV-17, XE-60, QF-1 (2:1:2) Diatomite CQ	ECD	Water	0.007mg/kg	134
Thiocarbamic acid herbicides	Silicone DS-550, Chromaton N-AW-DMCS	TID	Water	0.02mg/kg	135
Chlorobenzenes	Methyl phenyl silicone & Bentone 34	ECD	Waste water	0.02 - 150 ppb/200 ml	136
Monochlorophenoxy-alkane carboxylic acids	OV-17 on Chromaton N	ECD	Water	0.005mg/ℓ	137
Mirex + photomirex	SP2100 + SP-2401 on Supelcoport	ECD	Sediment	0.5mg/kg	138
Nitrophenol	Apiezon L + H_3PO_4 on Chromosorb W.	MS	Fresh water ocean sediment	0.17mg	139
Nitroxylenes	--	FID	Waste water	0.1g/ℓ	140
Phenols	OV-17 or OV-101 or 1% SP1240 DA	FID	Water	0.08μmole/ℓ	141
Styrene	Tenax - GC	MS	Water	0.1μg/ℓ	142
Polyhalogenated phenolic cpds.	4% SE-30 + 6% QF-1 on 80-100 mesh Supelcoport	^{3}H, ECD	Drinking water	10 ppt	143
Hydrocarbons	4mm x 120cm S.S. 7% Carbowax 20M on Chromosorb WAW	FID	Tap water	--	144
Chlorophenols	2% T-100 + 0.5% H_3PO_4 on Gas-Chrom Q	ECD, MS	Water	2μg/ℓ	145
Trace organics from manufacture of acetylene	10% PMS-100 polymethyl-siloxane on Celite 545. 20% PEGS	FID	Water	5mg/ℓ	146

TABLE 5.11 (Cont.)

Compound(s) type	Column(s)	Detector(s)	Sample type	Detection limit(s)	Reference
Pesticides	SE-30 on Chromosorb N-AW-DMCS	FID, ECD, FPD	Water	0.0005ng	147
Volatile halo-organics	3m Chromosorb 102	PID	Water	1µg/ℓ	148
Alkylbenzenes	Apiezon L	FID	Water	0.0006%	149
Organic Priority Pollutants (U. S. EPA). Volatile Organo-halides	1% SP-1000 on Carbopack B n-Octane on Porasil-C	Hall Model 700A conductivity	Water	0.0009 - 0.09µg/ℓ	150
Volatile aromatics	5% SP-2100 + 1.75% Bentone 34 on Supelcoport	HNU, PID	Water	Not determined	150
Acrolein and acrylonitrile	Chromosorb 101	FID	Water	2µg/ℓ	150
Phenols	1% SP-1240 DA on Supelcoport	FID	Water	1.4-10µg/ℓ	150
Phenols as pentafluorobenzyl derivatives	5% OV-17 on Chromosorb W-AW-DMCS	ECD	Water	--	150
Phthalate esters	1.5% SP2250 + 1.95% SP2401 on Supelcoport. 3% OV-1 on Supelcoport	ECD, FID	Water	0.02-0.13, 14-31µg/ℓ	150
Nitrosamines	10% Carbowax 20M + 2% KOH on on Chromosorb W-AW. 10% SP-2250 on Supelcoport	Nitrogen-Phosphorus Hall Reductive	Water	0.3-1.0µg/ℓ	150
Pesticides + polychlorinated biphenyls	1.5% SP2250/1.95% SP2401 on Supelcoport. 3% OV-1 on Supelcoport	ECD	Water	0.002-0.04µg/ℓ	150
Nitroaromatics + Isophorone	1.95% OF-1 + 1.5% OV-17 on Gas Chrom Q. 3% OV-101 on Gas Chrom Q	FID, ECD	Water	5µg/ℓ 0.06µg/ℓ	150
Haloethers	3% SP-1000 on Supelcoport Tenax-GC	Hall Conductivity Hall Conductivity	Water Water	0.04-2.2 µg/ℓ	150
Chlorinated hydrocarbons	1.5% OV-1 + 1.5% OV-225 on Gas Chrom Q	ECD	Water	0.001-0.018µg/ℓ	150

TABLE 5.11 (Cont.)

Compound(s) type	Column(s)	Detector(s)	Sample type	Detection limit(s)	Reference
Nitrite ion (diazotization)	1.5% OV-1 + 1.5% OV-225 on Gas Chrom Q	ECD	Water	0.5 ppb	151
Permethrin	--	Ion monitoring MS	Raw waste-water	0.05 ppb	152
209 PCB Isomers	--	MS	Commercial, environmental	0.01µg/ℓ	153
N-Nitrosamines	--	MS Chemilumine-scence	Sewage water	1µg/ℓ	154
Kepone	--	ECD, MS	Natural waters	2ng/ℓ	155
Volatile organics	--	FID, ECD, MS	Water supply	ng Range	156
Dinitrotoluene isomers	--	ECD	Sea water Industrial effluent	0.03µg/ℓ	157

REFERENCES

1 C.G. Creed, Res./Develop., 27 (9) (1976) 40.
2 G.A. Junk, J.J. Richard, M.D. Griesser, D. Witiak, J.L. Witiak, M.D. Arguello, R. Vick, H.J. Svec, J.S. Fritx and G.V. Calder, J. Chromatogr, 99 (1974) 745.
3 G.A. Junk, H.J. Svec, R.D. Vick and M.J. Avery, Environ. Sci. Technol., 8 (1974) 100.
4 B. Versino, H. Knöppel, M. DeGroat, A. Peil, J. Poelman, H. Schauenburg, H. Vissers and F. Geiss, J. Chromatogr., 122 (1976) 373.
5 P.R. Musty and G. Nickless, J. Chromatogr., 100 (1974) 83.
6 A.T. Borsocchi and R. Knobel, Am. Lab., 12 (2) (1980) 81.
7 R.S. Braman, in R.L. Grob, (Ed.), Chromatographic Analysis of the Environment, Dekker, New York, 1975, pp. 82.
8 J.E. Lovelock, in R.P. Scott, (Ed.), Gas Chromatography 1960, Butterworths, London, 1960.
9 D.H. Desty, C.J. Geach and A. Goldup, in R.P. Scott, (Ed.), Gas Chromatography 1960, Butterworths, London, 1960.
10 G.O. Nelson, Controlled Test Atmospheres, Principles and Techniques, Ann Arbor Press, Ann Arbor, MI., 1971, pp. 95-161.
11 J.M. McKelvey and H.E. Hoelscher, Anal. Chem., 29 (1957) 123.
12 A.P. Altshuller and I.R. Cohen, Anal. Chem., 32 (1960) 802.
13 R.L. Grob, Environmental Studies of the Atmosphere with Gas Chromatography, in D.M. Hercules, G.M. Hieftje, L.R. Snyder and M.A. Evenson, (Eds.), Contemporary Topics in Analytical and Clinical Chemistry, Plenum, New York, 1977, pp. 127-194.
14 B.E. Saltzman and C.A. Clemons, Anal. Chem., 38 (1966) 800.
15 B.E. Saltzman, C.A. Clemons, and A.E. Coleman, Anal. Chem., 38 (1966) 753.
16 F.H. Huyten, G.W.A. Rijinders and W.V. Beersum, in M. vanSwaay, (Ed.), Gas Chromatography 1962, Butterworths, London, 1963.
17 E.R. Stephens and M.H. Price, J. Air Pollut. Control Assoc., 15 (1965) 320.
18 G.F. Collins, F.F. Barlett, A. Turk, S.M. Edmonds and H.L. Mark, J. Air Pollut. Control Assoc., 15 (1965) 109.
19 C.C. Anderson, E.C. Gunderson, D.M. Coulson, B. Goodwin and K.T. Menzies, Generation of Test Atmospheres of Toxic Substances for Evaluation of Air Sampling Methods, in D.D. Dollberg and A.W. Verstuyft, (Eds.), Analytical

Techniques in Occupational Health Chemistry, ACS Symposium Series 120, American Chemical Society, Washington, DC, 1980, pp. 1-11.
20 A.E. O'Keefe and G.C. Ortman, Anal. Chem., 38 (1966) 760.
21 A.L. Lynch, R.F. Stalzer and D.T. Lefferts, Am. Ind. Hyg. Assoc. Q., 28 (1968) 79.
22 Methods of Air Sampling and Analysis, Intersociety Committee Publication, American Public Health Association, Washington, DC, (1972), pp. 22, 24, 29.
23 Analysis of Atmospheric Inorganics, U.S. Dept. of HEW, EPA, NAPCO, Cincinnati, Ohio, 1976, Section VII-2, pp. 1.
24 F.J. Debbrecht, in Modern Practice of Gas Chromatography, R.L. Grob, (Ed.), Wiley, New York, 1977, pp. 166-211.
25 J.E. Lovelock, Anal. Chem., 33 (1961) 162.
26 F. Bruner, C. Canulli and M. Passanzini, Anal. Chem., 45 (1973) 1790.
27 R.S. Braman and E.S. Gordon, IEEE Trans. Instrumental and Measurement, IM-14, (1965) 11-19.
28 T.A. Bellar, M.F. Brown and J.E. Sigsby, Jr., Anal. Chem., 35 (1963) 1924.
29 A.P. Altshuller and C.A. Clemons, Anal. Chem., 34 (1962) 466.
30 W. Leithe, The Analysis of Air Pollutants, Ann Arbor Science Publishers, Ann Arbor, MI, 1971 Chapter 2.
31 L. Angely, E. Levart, G. Guiochon, and G. Perslerbe, Anal. Chem., 41 (11) (1969) 1446.
32 W.R. Averett, J. Chromatogr. Sci., 8 (9) (1970) 552.
33 J.M. Brockway, A.W. Boyne and J.G. Gordon, J. Appl. Physiol., 31 (2) (1971) 296.
34 H.D. Axelrod, R.J. Teck, J.J. Lodge and R.H. Allen, Anal. Chem., 43 (3) (1971) 496.
35 Y. Hashimoto and S. Tanaka, Proceedings of the 4th Int. Clean Air Congress, Tokyo, May 16-20, 1977, pp. 349-351.
36 O.F. Folmer, Anal. Chim. Acta, 56 (3) (1971) 440.
37 F.J. Debbrecht, in R.L. Grob, (Ed.), Modern Practice of Gas Chromatography, Wiley, New York, 1977, pp. 203-210.
38 R. Schill, in R.L. Grob, (Ed.), Modern Practice of Gas Chromatography, Wiley, New York, 1977, pp. 304-321.
39 J.C. Cavagnol and W.R. Betker, L.S. Ettre and A. Zlatkis, (Eds.), The Practice of Gas Chromatography, Wiley-Interscience, New York, 1967, pp. 106-123.
40 E.L. Szonntagh, in L.S. Ettre and A. Zlatkis, (Eds.), The Practice of Gas Chromatography, Wiley-Interscience, New York, 1967, pp. 512-519.
41 J. Lebbe, in J. Tranchant, (Ed.), Practical Manual of Gas Chromatography, Elsevier, Amsterdam, 1969, pp. 54-62.
42 P.G. Jeffery and P.J. Kipping, Sample Transfer Systems, Macmillan, New York, 1964, Chapter 2.
43 G.L. Pratt and J.H. Purnell, Anal. Chem., 32 (1960) 1213.
44 D.N. Glew and D.M. Young, Anal. Chem., 30 (1958) 1890.
45 D.S. Russell and M.E. Bednas, Anal. Chem., 29 (1957) 1562.
45A M.A. Kaiser and B.J. Wood, J. Chromatog. Sci., 19 (1981) 104.
46 L. Rohrschneider, J. Chromatogr., 22 (1966) 6.
47 L. Rohrschneider, Adv. Chromatogr., 4 (1967) 333.
48 W.O. McReynolds, J. Chromatog. Sci., 8 (1970) 685.
49 W.R. Supina, The Packed Column in Gas Chromatography, Supelco, Bellefonte, PA, 1974.
50 K. Grob and G. Grob, Identification and Analysis of Organic Pollutants in Water, Ann Arbor Science, Ann Arbor, MI, 1976, pp. 75.
51 L.S. Ettre, Open Tubular Columns in Gas Chromatography, Plenum, New York, 1965.
52 G. Schomberg, Gaschromatographie, Verlag Chemie GmbH, Weinheim, 1977.
53 R.R. Freeman, (Ed.), High Resolution Gas Chromatography, Hewlett Packard Co., Avondale, PA, 1979.
54 W. Jennings, Gas Chromatography with Glass Capillary Columns, 2nd. ed., Academic Press, New York, 1980.
54A K. Grob and K. Grob, Jr., J. Chromatogr., 151 (1978) 311.
54B M. Galli, S. Trestianu and K. Grob, Jr., HRC & CC, 2(1979) 366-370.
54C M. Galli and S. Trestianu, J. Chromatogr., 203(1981) 193-205.
55 D.J. David, Gas Chromatographic Detectors, Wiley-Interscience, New York, 1974.

56 J. Sevcik, Detectors in Gas Chromatography, Elsevier, Amsterdam, 1976.
57 J.J. Sullivan and M.J. O'Brien, in R.L. Grob, (Ed.), Modern Practice of Gas Chromatography, Wiley, New York, 1977, pp. 213-288.
58 W. McFadden, Techniques of Combined Gas Chromatography/Mass Spectrometry, Wiley-Interscience, New York, 1973.
59 E. Kovats, Helv. Chim. Acta, 41 (1958) 1915.
60 A. Wehrli and E. Kovats, Helv. Chim. Acta, 42 (1959) 2709.
61 E. Kovats, Z. Anal. Chem., 181 (1961) 351.
62 E. Kovats, Helv. Chim. Acta, 46 (1963) 2705.
63 K.P. Hupe, J. Gas Chromatog., 3 (1965) 12.
64 G. Guiochon, Anal. Chem., 36 (1964) 1672.
65 B.A. Galushkin, V.D. Grishin, Zh. Anal. Khim., 35 (1980) 2039.
66 L.R. Snyder and J.J. Kirkland, Introduction to Modern Liquid Chromatography, Wiley-Interscience, New York, 1979.
67 J.C. Touchstone and M.F. Dobbins, Practice of Thin Layer Chromatography, Wiley, New York, 1978.
68 J. Janak, I. Klimes and K. Hana, J. Chromatogr., 18 (1965) 270.
69 G.V. Golovkin, A.I. Smol'chenko, L.N. Zotov, S.E. Sazonov, Zh. Anal. Khim., 35 (1980) 964; C.A., 93 (1980) 63372p.
70 E.B. Ledford, Jr., R.L. White, S. Ghadiri, C.L. Wilkins and M.L. Gross, Anal. Chem., 52 (1980) 2450.
71 R.K. Mitchum, G.F. Moler and W.A. Korfmacher, Anal. Chem., 52 (1980) 2278.
72 F.L. Onuska, M.E. Comba and J.S. Coburn, Anal. Chem., 52 (1980) 2272.
73 J. Andrzej, M. Koralewska, B. Zielinska and M. Sochacki, Kohs. Smola. Gaz, 25 (1980) 95; C.A. 93 (1980) 222940r.
74 A. Baranski, S. Ceekiewicz and J. Galuszka, Wiss. Z.-Friedrich-Schiller-Univ. Jena, Math.-Naturwiss. Reihe, 27 (1978) 595; C.A. 93 (1980) 226312k.
75 R. Vistocco and P. Beggio, AES, 2 (1980) 51; C.A. 93 (1980) 209507k.
76 D.R. Gage and S.O. Farwell, Anal. Chem., 52 (1980) 2422.
77 F. Chow and A. Karmen, Clin. Chem., 26 (1980) 1480.
78 V. Felt and P. Husek, J. Chromatogr., 197 (1980) 226.
79 A. Pannatier and B. Testa, Pharm. Acta Helv., 55 (1980) 100.
80 Y.G. Kessel'man and I.P. Ogloblina, Khim. Prom-st., Ser.: Khlornaya Prom-st., (1) (1980) 9; C.A. 93 (1980) 215054b.
81 F. Chou, Wei Sheng Wu Hsueh T'ung Pao, 7 (3) (1980) 136; C.A. 93 (1980) 217869v.
82 F.J. Debbrecht, Part II, Quantitative Analysis, in R.L. Grob, (Ed.), Modern, Practice of Gas Chromatography, Wiley, New York, 1977, pp. 166.
83 J. Novak, Quantitative Analysis by Gas Chromatography, Dekker, New York, 1975.
84 I.G. Young, Amer. Lab., 7 (1975) 27; 7 (1975) 37; 7 (1975) 11.
85 D.M. Rosie and R.L. Grob, Anal. Chem., 29 (1957) 1263.
86 Application Reviews, Anal. Chem., 47 (1975) 1R-364R; 49 (1977) 1R-286R; 51 (1979) 1R-342R.
87 Fundamental Reviews, Anal. Chem., 48 (1976) 1R-442R; 50 (1978) 1R-384R; 52 (1980) 1R-383R.
88 L.L. Lamparski and T.J. Nestrick, Anal. Chem., 52 (1980) 2045.
89 K.D. Brunnemann, W. Fink and F. Moser, Oncology, 37 (1980) 217.
90 M.A. Klisenko and D.B. Girenko, Gig. Sanit., (9) (1980) 57; C.A. 93 (1980) 216149y.
91 R.F. Arrendale, R.F. Severson and M.E. Snook, Beitr. Tobakforsch. Int., 10 (1980) 100; C.A. 93 (1980) 218116j.
92 J.B. Mann, J.J. Freal, H.F. Enos and J.X. Danauskas, J. Environ. Sci. Health, B15 (1980) 507.
93 J.B. Mann, J.J. Freal, H.F. Enos and J.X. Danauskas, J. Environ. Sci. Health, B15 (1980) 519.
94 Y. Saint-Jalm and P. Mooree-Testa, J. Chromatogr., 198 (1980) 188.
95 K. Funazo, M. Tanaka and T. Shono, Anal. Chim. Acta, 119 (1980) 291.
96 L.D. Molchanov, N.F. Rogozhkina and V.K. Golyshkin, Khim. Prom-st., Ser.:Metody Anal. Kontrolya Kach. Prod. Khim. Prom-sti., (7) (1980) 7; C.A. 93 (1980) 215066g.
97 E. Matisova, J. Krupcik and O. Liska, J. Chromatogr., 173 (1979) 139.

98 J.S. Hobbs, Int. Environ. Saf., March-April (1978) 22.
99 X. Guardino, J. Bartual and M.J. Berenguer, Afinidad, 36 (1979) 9; C.A. 91 (1979) 61768g.
100 M. Paabo, M.M. Birky and S.W. Womble, J. Combust. Toxicol., 6 (1979) 99.
101 K. Shimohara, S. Sueta, T. Tabata and N. Shigemori, Taiki Osen Gakkaishi, 14 (1979) 31; C.A. 91 (1979) 128153y.
102 J. Kasche and K. Schroeter, Chem. Tech. (Leipzig), 31 (1979) 311; C.A. 91 (1979) 128246f.
103 H.J. Rath, D. Schmidt and J. Wimmer, Chromatographia, 12 (1979) 567.
104 K.K. Choi and K.W. Fung, Analyst (London), 104 (1979) 455.
105 W.S. Schlotzhauer, O.T. Chortyk and R.F. Severson, Tob. Int., 181 (1979) 63.
106 N. Fukuda, H. Itoh, A. Tsukamoto and H. Tamari, Bunseki Kagaku, 28 (1979) 569; C.A. 92 (1980) 27645a.
107 A. Dahms and W. Metzner, Holz Roh-Werkst., 37 (1979) 341; C.A. 92 (1980) 35229h.
108 B.V. Koval'chuk, N.G. Kolesnik and L.F. Pushkareva, Khim. Prom-st., Ser.:Metody Anal. Kontrolya Kach. Prod. Khim. Prom-sti., (7) (1979) 1; C.A. 92 (1980) 81361t.
109 E.B. Koroleva, L.V. Sakharova, A.L. Kondrasheva and L.B. Sevryugov, Gig. Sanit., (11) (1979) 67; C.A. 92 (1980) 81372x.
110 P.C. Ensor, J. Assoc. Public Anal., 16 (1978) 129; C.A. 92 (1980) 98661p.
111 I. Kifune, K. Kawata and K. Oikawa, Eisei Kagaku, 25 (1979) 205; C.A. 92 (1980) 115497j.
112 V.I. Lyashenko and V.E. Prisyazhnyuk, Gig. Sanit., (11) (1979) 56; C.A. 92 (1980) 152138p.
113 V.B. Dorogova, L.I. Zuitseva and V.A. Khomutova, Gig. Tr. Prof. Zabol., (12) (1979) 51; C.A. 92 (1980) 184961t.
114 A. Krynska and M. Posniak, Pr. Cent. Inst. Ochr. Pr., 28 (1978) 389; C.A. 92 (1980) 202641v.
115 R.E. Clement, G.A. Eiceman, F.W. Karasek, W.D. Bowers and M.L. Parsons, J. Chromatogr., 189 (1980) 53.
116 M. Nagase, Bunseki Kagaku Sap. Anal. 29 (1980) 293; C.A. 93 (1980) 100717c.
117 T.N. Shakina, A.I. Kuznetsova and L.F. Abramas, Khim. Prom-st., Ser.:Metody Anal. Kontrolya Kach. Prod. Khim. Prom-sti., (4) (1980) 10; C.A. 93 (1980) 106535p.
118 K. Watabe, S. Iida, T. Ichikawa and M. Kondo, Eisei Kagaku, 25 (1979) 343; C.A. 93 (1980) 154984t.
119 W.R.A. DeJonghe, D. Chakaborti and F.C. Adams, Anal. Chem., 52 (1980) 1974.
120 K. Ono and T. Hayakawa, Taiki Osen Gakkaishi, 14 (1979) 479; C.A. 93 (1980) 244500z.
121 H. Kachi and Y. Kakano, Jidosha Gijutsukai, 20 (1980) 64; C.A. 93 (1980) 244576d.
122 B.G. Oliver and K.B. Bothen, Anal. Chem., 52 (1980) 2066.
123 R.S. Narang and B. Bush, Anal. Chem., 52 (1980) 2076.
124 P. Buryan, J. Riha, A. Capek and J. Macak, Plyn., 60 (1980) 6; C.A. 93 (1980) 209663h.
125 D. Barcelo, and I. Eek, Pergamon Ser. Environ. Sci., 3 (1980) 335; C.A. 93 (1980) 209711x.
126 T. Tsukioka, M. Maruyama and T. Katsuno, Nagano-ken Eisei Kogai Kenkyusho Kenkyu Hokoku, 2 (1980) 86; C.A. 93 (1980) 215057e.
127 P.W. Albro and C.E. Parker, J. Chromatogr., 169 (1979) 161.
128 K. Kenmotsu, K. Matsunaga and T. Ishida, Okayama-ken Kankyo Hoken Senta Nempo, 2 (1978) 164; C.A. 91 (1979) 32413g.
129 A. Lablanche, B. Soulard, G. Martin and E. David, Tech. Eau Assainissement, 387 (1979) 17; C.A. 91 (1979) 33956y.
130 G. Brozowski, D. Burkitt, M. Gabriel, J. Hanrahan, E. McCarthy and J. Smith, NBS Spec. Publ., Washington, DC, 519 (1979) 175.
131 L.M. Neretina and L.Y. Yarpovetskii, Khim. Tekhnol. (Kiev), (3) (1979) 19; C.A. 91 (1979) 145490y.
132 Organochlorine Insecticides and Polychlorinated Biphenyls in Water, United Kingdom Dept. of the Environment Publ., National Water Council, London, 1979.

133 C.W. Luhning, P.D. Harman, J.B. Sills, V.K. Dawson and J.L. Allen, J. Assoc. Off. Anal. Chem., 62 (1979) 1141.
134 H. Dubsky, K. Hana, M. Komarkova and B. Rittich, Vet. Med. (Prague), 24 (1979) 493; C.A. 92 (1980) 4809t.
135 I.S. Kofman and V.I. Kofanov, Gig. Sanit., (10) (1979) 41; C.A. 92 (1980) 35763c.
136 Z.C. Bao and C.H. Chao, Huan Ching K'o Hsueh, (6) (1979) 50; C.A. 92 (1980) 99353b.
137 V.D. Chmil, Zh. Anal. Khim., 35 (1980) 132; C.A. 92 (1980) 123199d.
138 R.J. Norstrom, H.T. Won, M. Van Hove Holdrinet, P.G. Calway and C.D. Naftel, J. Assoc. Off. Anal. Chem., 63 (1980) 37.
139 K. Matsunaga, F. Shibata, N. Saito, N. Torigashi, K. Kenmochi and Y. Ogino, Okayama-ken Kankyo Hoken Senta Nempo, 3 (1979) 192; C.A. 92 (1980) 140189t.
140 L.A. Shutova, N.M. Stupina, V.V. Zhdanova and G.M. Permyakova, Khim. Prom-st., Ser.:Metody Anal. Kontrolya Kach. Prod. Khim. Prom-sti., (8) (1979) 7; C.A. 92 (1980) 168552e.
141 R.T. Coutts, E.E. Hargesheimer and F.M. Pasutto, J. Chromatogr., 179 (1979) 291.
142 B.I. Brookes, S.M. Kickells and R.S. Nicolson, J. Assoc. Public Anal., 16 (1978) 117; C.A. 92 (1980) 168923b.
143 C. Morgade, A. Barquet and C.D. Pfaffenberger, Bull. Environ. Contam. Toxicol., 24 (1980) 257.
144 V. Durdovic, Acta Hydrochim. Hydrobiol., 7 (1979) 527; C.A. 92 (1980) 185573e.
145 T. Kodama, Aichi-ken Kogai Chosa Senta Shoho, 7(1979) 8; C.A. 92 (1980) 208565v.
146 L.S. Masalova, N.N. Kedrina and V.I. Stavrati, Khim. Prom-st., Ser.:Metody Anal. Kontrolya Kach. Prod. Khim. Prom-sti., (6) (1980) 21; C.A. 93 (1980) 142373t.
147 E.I. Babkina, A.V. Debtseva and A.A. Siverina, Khim. Sel'sk. Khoz. 18 (1980) 55; C.A. 93 (1980) 162521t.
148 R.S. Narang and B. Bush, Anal. Chem., 52 (1980) 2076.
149 A.D. Pakhomova, V.L. Berendeeva, L.A. Kalinichenko and I.T. Goronovskii, Khim. Tekhnol. Vody, 2 (1980) 33.
150 Federal Register, Part III, 44 (1979) 69464, U.S. Federal Government Printing Office, Washington, DC.
151 T. Chikamoto and S. Nagata, Chem. Lett., (6) (1980) 737.
152 M.M. Siegel, B.E. Hildebrand and D.R. Hall, Int. J. Environ. Anal. Chem., 8 (1980) 107.
153 G.W. Tindall and P.E. Wininger, J. Chromatogr., 196 (1980) 109.
154 M.L. Richardson, K.S. Webb and T.A. Gough, Exotoxicol. Environ. Saf., 4 (1980) 207.
155 I.R. DeLeon, P.C. Remele, S.K. Miles and J.L. Laseter, Anal. Lett., 13 (1980) 504.
156 A.P. Sandhofer, Swiss Chem., 2 (1980) 41; C.A. 93 (1980) 245130r.
157 A. Hashimoto, H. Sakino, E. Yamagami and S. Tateishi, Analyst (London), 105 (1980) 787.

RELATED READINGS ON AIR AND WATER ANALYSES

1 Gas Chromatographic Method for the Determination of Traces of Carbon Disulfide in the Air. E. Saulova, M. Pankova, Khim. Ind. (Sofia), (4)(1980)164.

2 Application of a Direct Aqueous Acetylation Technique to the Gas Chromatographic Quantitation of Nitrophenols and 1-Naphthol in Environmental Water Samples. R.T. Coutts, E.E. Hargesheimer, F.M. Pasutto, J. Chromatogr., 195(1980)105.

3 Methods of Evaluation of Potentially Hazardous Organics in Water. R.P. Collins, Report 1980, W80-02761, OWRT-A-072-CONN(1); Inst. Water Resourc., Univ. Connecticut, Storrs, CT, USA.

4 The Analysis of Organic Water Pollutants by Gas Chromatography and Gas Chromatography-Mass Spectrometry. R.A. Hites, Adv. Chromatogr., 15(1977) 69-112, Dekker, N.Y.

5 Determination of Polybrominated Biphenyl and Related Compounds by Gas-Liquid Chromatography with a Plasma Emission Detector. K.J. Mulligan, J.A. Caruso, F.L. Fricke, Analyst (London), 105(1980)1060.

6 Problems and Pitfalls in Acetaldehyde Determinations. C.J.P. Eriksson, Alcohol.: Clin. Exp. Res., 4(1)(1980)22-9.

7 Sensitive Gas Chromatographic Determination of Phenols as Bromophenols using Electron Capture Detection. Y. Hoshika, G. Muto, J. Chromatogr., 179 (1)(1979)105-11.

8 Determination of Vapors of Organic Solvents in the Working Atmosphere by Capillary Gas Chromatography. J. Hrivnak, M. Revus, Acta Fac. Rerum Nat. Univ. Comenianae, Form. Prot. Nat., 4(1979)63-72.

9 Analytical Method of C_5-C_{20} Hydrocarbons in Automotive Exhaust by Glass Capillary Columns. T. Itoh, H. Kachi, Y. Nakano, Jidosha Gijutsukai Ronbunshu, 19(1979)11-16.

10 Development and Application of an Automated Airborne Methane-Nonmethane Hydrocarbon Analyzer. W.R. Cofer III, Proc., Annu. Meet. - Air Pollut. Control Assoc., 72(4)(1979).

11 Analyses of Water Pollutants. 5. Analysis of Soap and Synthetic Cleansing Agents. Y. Ogino, M. Nagao, Okayama-ken Kankyo Hoken Sento Nempo, 3(1979) 283-6.

12 Spectrometric and Chromatographic Methods in Water Analysis. G. Franke, H. Hein, Chem.-Tech. (Heidelberg), 8(6)(1979)295-302.

13 Analyses of Marine Sediments. 3. Determinations of Paraffinic Hydrocarbons and Organic Sulfur. M. Imanaka, K. Matsunaga, H. Hata and T. Ishida, Okayama-ken Kankyo Hoken Senta Nempo, 3(1979)131-41.

14 Identification and Analysis of Organic Substances in the Environment. Analyses of Trace Organics in Water and Sediment. M. Koga, R. Shinohara, A. Kido and S. Etoh, Koenshu-Iyo Masu Kenkyukai, 4(1979)153-9.

15 Determination of the Concentration of Physiological and Volatile Anesthetic Gases in Respiratory Gases by Gas Chromatography. F. Zheng, Yu-Mei Xie and Li-De Jin, Chung-hua I Huseh Tsa Chih, 59(8)(1979)495-9.

16 Glass Capillary Gas Chromatography of Chlorinated Dibenzofurans, Chlorinated Anisoles and Brominated Biphenyls. T.J. Farrell, J. Chromatogr. Sci., 18(1)(1980)10-17.

17 A Selective GC Column for Trace Analysis of Vinyl Chloride in Air. D. Krockenberger, H. Lorkowski and L. Rohrschneider, Chromatographia, 12(12) (1979)787-9.

18 Analyses of Micro Organic Substances in Tap Water by GC-MS-COM System, High Resolution-GC-MS and Mass Fragmentography. R. Shinohara, M. Koga, A. Kido, S. Eto, T. Hori and T. Akiyama, Koenshu-Iyo Masu Kenkyukai, 3 (1978)69-76.

19 Selection and Chromatographic Analysis of Gas Samples in the Vacuum Treatment of Molten Steel. Y.T. Krut, I.I. Ansheles, V.L. Safonov, A.A. Ezhov, S.I. D'yakov and G.F. Yakovenko, Zabod. Lab., 46(1)(1980)11-12.

20 Gas-Chromatographic Determination of Hydrogen Cyanide and Dicyanogen. V.K. Steba, V.I. Zrazhevskii and V.D. Parkhomenko, Metody Anal. Kontrolya Kach. Prod. Khim. Prom. Sti., (11)(1979)6-8.

21 Column Deactivation for the Analysis of Picogram Amounts of Underivatized Chlorophenols and Chlorocresols by Combined Gas Chromatography`Mass Spectrometry. M.A. White and K.R. Parsley, Biomed Mass Spectrom., 6(12)(1979) 570-2.
22 Organic Matter of Groundwater According to Data of Gas-Chromatography Studies. S.G. Mel'kanovitskaya, Geokhimiya, (2)(1980)272-85.
23 Application of Head-Space Analysis to the Study of Volatile Organic Impurity Concentrates. Gas Chromatographic Determination of Aromatic Hydrocarbons in Atmospheric Air. B.V. Ioffe, A.G. Vitenberg and I.A. Isibul'skaya, J. Chromatogr., 186(1979)851-9.
24 Gas Chromatographic Profile Analysis of Polycyclic Aromatic Hydrocarbons in Water. G. Grimmer and K.W. Naujack, Vom Wasser, 53(1979)1-8.
25 Gas Chromatographic Determination of DDT in Industrial Wastewater and Environmental Water with Stationary Liquids and Supports. Ju-Si Wang and L.H. Zhao, Huan Ching K'o Hsueh, 1(1)(1980)20-5.
26 Gas Chromatographic Method for the Analysis of Polychlorinated Biphenyls in Transformer Oil. J. Ogata, J.D. Okun, J.W. Hylin and A. Bevenue, J. Chromatogr., 189(3)(1980)425-7.
27 An Air Sampling and Analysis Method for Monitoring Personal Exposure to Vapors of Acrylate Monomers. M.W. Bosserman and N.H. Ketcham, Am. Ind. Hyg. Assoc. J., 41(1)(1980)20-6.
28 Gas-Chromatographic Determination of Hydrocarbons in extracts of Atmospheric Dust. L.S. Lysyuk and A.N. Korol, Sorbenty Dlya Gaz. i Zhidkostnoi Khromatogr., M (1979)82-5.
29 Direct Aqueous Injection Gas Chromatography for the Analysis of Trace Organics in Water. P. Simmonds and E. Kerns, J. Chromatogr., 186(1979)785-94.
30 The Concentration, Isolation, and Determination of Acidic Material from Aqueous Solution. J.J. Richard and J.S. Fritz, J. Chromatogr., Sci., 18(1)(1980)35-8.
31 Determination of Ammonium-Nitrogen in Water. T. Ito, Jpn. Kokai Tokkyo Koho 80 01,552 (Cl. G01N31/00), 08 Jan 1980.
32 Steam-Modified Gas-Solid Chromatography: a Complementary Technique for Organic Pollutant Survey. C.L. Guillemin, F. Martinez and S. Thiault, J. Chromatogr. Sci., 17(19)(1979)677-81.
33 Chromatographic Analysis of Major Trace Components in Solvents with Concentration Prior to Sample Injection. V.G. Berezkin, M.N. Budantseva and E. Dows, J. Chromatogr., 191(1)(1980)309-12.
34 Measurement of Petroleum Oil Deposits on Citrus Leaves by Gas-Liquid Chromatography. G.O. Furness, R.T. Howard and P.J. Smith, Pestic. Sci., 10(6) (1979)478-84.
35 Simple Multi-Residue Analysis of Organophosphorus Pesticides by Mass Chromatography. H. Kobayashi, M. Kojima, O. Matano and S. Goto, Nippon Noyaku Gakkaishi, 4(4)(1979)463-72.
36 Determination of Carbofuran Residues in Potato, Onion and Turnip; Comparative Analyses Using Gas Chromatography-Selected Ion Monitoring or High-Performance Liquid Chromatography. J.R. Robinson and R.A. Chapman, J. Chromatogr., 193(2)(1980)213-24.
37 Analytical Method for Residues of Dithiocarbamate Fungicides in Agricultural Products. M. Uno, T. Okada, Y. Onji, T. Ohmae and Y. Nishikawa, Shokunin Eiseigaku Zasshi, 20(6)(1979)450-5.
38 Activated Carbon for Analysis. S. Kuendig and D. Bauer, Chem. Rundsch., 33(8)(1980)1,6,9.
39 Quantitative Treatment of Chromatograms of Coke-Oven Gas and Roasting-Sintering Gas. M.N. Borisova, N.D. Afanas'ev, Y.G. Musatova and V.A. Botryakova, Khim. Prom-st., Ser.: Fosfornaya Prom-st., (6)(1979)49-52.
40 Determination of Residual 2,6-Dichlorobenzonitrile in Soil and Water. F. Herzel, J. Chromatogr., 193(2)(1980)320-1.
41 Study of Pesticide Residues in Strawberry. G. Visi and F. Orosz, Novenyvedelem (Budapest), 14(19)(1978)408-11.

42 Organic Solvent-Soluble Organic Matter from Soils Underlying Native Range and Crested Wheatgrass in Southeastern Alberta, Canada. J.F. Dormaar, A. Johnston and S. Smoliak, J. Range Manage., 33(2)(1980)99-101.
43 Gas Chromatograph for Problems of Environmental Protection. P. Popp, G. Oppermann and G. Andrae, ZfI-Mitt, 22, Suppl. (1979)94-105.
44 Microcoulometric Determination of Chlorinated Organic Compounds in Rhine Water. L. Braunstein, K. Hochmueller and K. Spengler, Vom Wasser, 53 (1979)27-51.
45 Analysis of Alkylated Mixtures of Polychlorinated Biphenyls by Capillary Gas Chromatography-Mass Spectrometry. J. Krupcik, P.A. Leclercq, J. Garaj and A. Simova, J. Chromatogr., 191(1)(1980)207-20.
46 High Resolution Gas Chromatography: An Overview. S.P. Cram and F.J. Yang, Adv. Chem. Ser., 185(1980)105-21.
47 Gas-Liquid Chromatographic Determination of Organophosphorus Insecticide Residues in Fruits and Vegetables. J.R. Ferreira and A.M.S. Fernandes, J. Assoc. Off. Anal. Chem., 63(3)(1980)517-22.
48 Modification of the AOAC Method for Determination of Fumigants in Wheat. M. Clower, Jr., J. Assoc. Off. Anal. Chem., 63(3)(1980)539-45.
49 Gas Chromatographic Analysis of Volatile Nitrogen Bases of Boiled Beef as Possible Precursors of N-Nitrosamines. R.V. Golovnya, I.L. Zhuravleva and Y.P. Kapustin, Chem. Senses Flavour, 4(2)(1979)97-105.
50 Gas-Chromatographic Determination of Carbon Monoxide in the Air. M.T. Dmitriev and G.M. Kolesnikov, Gig. Sanit., (3)(1980)53-4.
51 Analysis of Maleic, Fumaric, and Succinic Acids in Air by Use of Gas Chromatography. B.M. Wathne, Analyst (London), 105(1249)(1980)400-3.
52 Gas Chromatographic Determination of Carbon Monoxide in Atmospheric Air. N.S. Nosova and N.N. Tomina, Sb. Nauch. Tr. Mosk. NII Gigieny, (13)(1979) 63-4.
53 Gas-Chromatographic Procedure for the Measurement of Concentrations of Methyl, Ethyl, Isopropyl, and N-Butyl Alcohols in the Air of Industrial Plants. E.K. Prokhorova, V.E. Kalinin, V.L. Boyarskova and V.N. Blinov, Kompleks, Probl. Okhrany Truda, M. (1979)116-19.
54 Studies on the Sulfur Compounds in Crude Oils and Heavy Fuel Oils Exposed on the Sea Surface by the FPD Gas Chromatography. 2. S. Utashiro and H. Matsuo., Kenkyu Holoku - Kaijo Hoan Daigakko, Dai-2-bu, 25(1)(1979)11-21.
55 Gas-Liquid Chromatography Applied to the Analysis of Complex Environmental Samples. A Preliminary Investigation of a Method of Simultaneous Extraction and Derivatization of Solid Samples before Analysis by Gas-Liquid Chromatography. A. Waggott and C.L. Saunders, Tech. Rep. TR-Water Res. Cent., TR 44(1977)12 pp.
56 Determination of Selected Volatile Organic Priority Pollutants in Water by Computerized Gas Chromatography-Quadrupole Mass Spectrometry. W. Pereira and B.A. Hughes, J. Am. Water Works Assoc., 72(4)(1980)220-30.
57 A Simple and Sensitive Detection Method for Nitrilotriacetic Acid in Drinking and River Water. J.K. Reichert and A.H.M. Linckens, Environ. Technol. Lett., 1(1)(1980)42-9.
58 Microanalysis of Aqueous Samples for Phenols and Organic Acids. W.A. Prater, M.S. Simmons and K.H. Mancy, Anal. Lett., 13(A3)(1980)205-12.
59 A Study of the Partitioning of Phenols by Use of Spectroscopy and Gas Chromatography. R.L. Grob and M.L. Fultz, J. Environ. Sci. Health, Part A, A15(1)(1980)45-64.
60 Analysis of Polychlorinated Biphenyls (PCB) by Glass Capillary Gas Chromatography. Composition of Technical Aroclor-and Clophen-PCB Mixtures. K. Ballschmiter and M. Zell, Fresenius' Z. Anal. Chem., 302(1)(1980)20-31.
61 Ov-17-QF-1 Capillary Column for Organochlorine Pesticide Analysis. M.Cooke and A.G. Ober, J. Chromatogr., 195(2)(1980)265-9.
62 Evaluation of the Effectiveness of Fireproofing Agents by a Gas-Liquid Chromatographic Method. V. Anzor'e, N.I. Lytkina, L.N. Vansyatskaya and A.V. Zhurko, Kauch. Rezina, (7)(1980)55-6.

63 Gas-Chromatographic Determination of Small Concentrations of Dichloromethane in Purified Wastewater from Synthetic Fiber Production. N.M. Sergeeva., Khim. Volokna (1)(1980)54-5.

64 Nonextractive Gas Chromatographic Determination of Low Concentrations or Organic Substances Dissolved in Water. V. Y. Botsan and V.I. Kofanov, Khimiya i Tekhnol. Vody, 1(1)(1979)39-42.

65 Determination of Trichlorosilane and Silicon Tetrachloride in Hydrogen Chloride Exit Gas. S. Semenikhina, N.A.Palamarchuk, L.A. Nechaeva and B.P. Krasnov, Khim. Prom-st., Ser.: Metody Anal. Kontrolya Kach. Prod. Khim. Prom-sti., (4)(1980)24-6.

66 Chromatographic Determination of Polycyclic Aromatic Hydrocarbons. V.D. Ovsyanik, L.I. Uspanova and V.G. Matvienko, Khim. Prom-st., Ser.: Metody Anal. Kontrolya Kach. Prod. Khim. Prom-sti., (4)(1980)22-3.

67 Quantitative Determination of Trace Impurities of Lower Alcohols in Aqueous Solutions by Gas Chromatography. A.N. Korol, G.V. Filonenko, N. Omel'yanets and L.I. Petrik, Otkrytiya, Izobret., Prom. Obraztsy, Tovarnye Znaki, (20)(1980)218.

68 Determination of Formaldehyde in Room Atmosphere by Headspace Gas Chromatography. A. Kettrup, H. Stenner, J. Heuel and W. Lorek, Appl. Headspace Gas Chromatogr., (GC Headspace Symp.)(1980)23-31.

69 New Possibility of Monitoring Gas Emissions with an Ionization Detector. P. Popp, G. Oppermann and G. Andrae, ZfI-Mitt., 22(1979)154-64.

70 A Sampling and Analytical Method for Vinyl Acetate. D.L. Foerst and A.W. Teass, ACS Symp. Ser. 120(1980)169-84.

71 Gas-Chromatographic Procedure for Measuring Concentrations of Methyl, Ethyl, Isopropyl, and n-Butyl Alcohols in the Air of Industrial Buildings. E.K. Prokhorova, V.E. Kalinin, V.L. Boyarskova and V.N. Blinov, Compleks. Probl. Okhrany Truda, M., (1979)116-19.

72 Reduced Pressure Headspace Analysis (RPHA) for the Determination of Volatile Components in the Lower and Sub-ppm Range. W. Stotzka, Appl. Headspace Gas Chromatogr., (GC Headspace Symp.)(1978)(Pub.1980)56-60.

73 Detection of Environmental Pollutants. Method for Analyzing for Aromatic Nitro Compounds. H. Sugiyama, Y.Fushiwaki and K. Tanaka, Kanagawa-ken Kogai Senta Nempo, 10(1977)192.

74 GC-MS Determination of Volatile Organic Compounds in Water. D.P. Beggs, Oesterr. Chem. Z., 81(4)(1980)98-9.

75 Characterization of Chlorinated Hydrocarbons in the Marine Environment: Sampling and Analytical Aspects. S.P. Pavlou, R.N. Dexter and W. Hom, Pergamon Ser. Environ. Sci., 3(1980)185-96.

76 Interface for Capillary Columns on an Automatic Headspace Gas Chromatographic System. T. Meyer, HRC CC, J. High Resolut. Chromatogr. Chromatogr. Commun., 3(7)(1980)357-8.

77 Simultaneous Analysis of Respiratory Gases and Nitrous Oxide in Dental Patients. J.E. O'Reilly, G.I. Roth, J.L. Matheny and J.E. Dean, J. Dent. Res., 59(4)(1980)675-82.

78 Chemical Composition of Dissolved Gases of Subsurface Waters in Dagestan according to the Results of Gas-Chromatographic Analyses. Z.M. Shtanchaeva, I.D. Aziev and O.A. Mamaev, Akad. Nauk SSSR, Dagest. Fil., Inst. Geol. (Tr), 1(14)(1977)94-100.

79 Capillary Gas Chromatography Determination of C_2 to C_{11} Hydrocarbons in Ambient Air. I. A Statistical Estimation of Photochemical Reaction for Olefins and Aromatics. Y. Tsujino, Taiki Osen Gakkaishi, 14(6)(1979)231-42.

80 Chromatographic Methods of Analysis for Methanol and Ethanol in Automotive Exhaust. C.J. Raible and F.W. Cox, Report 1979, CONF-7906132-2, Energy Res. Abstr., 5(1)(1980).

81 Determination of Oxygen Content in the Air of an Industrial Type of Stock Farm by a Gas Chromatographic Method. T.A. Pak, V.I. Balanin, Y.V. Efimov and A.A. Khomchenko, Sb. Nauch. Tr. Leningr. Vet. In-t, 59(1979)18-24.

82 Determination of Perchloroethylene in the Sub-Parts-per-Billion Range in Ambient Air by Gas Chromatography with Electron Capture Detection. A.L. Sykes, D.E. Wagoner and C.E. Decker, Anal. Chem.,52(11)(1980)1630-3.

83 Determination of Lower Fatty Acids C_2-C_6 in Waste Waters and Sludges by Gas Chromatography. J. Sabata, J. Zabranska, J. Kriz and P. Grau, Sb. Vys. Sk. Chem.-Technol. Praze (Oddil)F: Technol. Vody Prostredi, F23(1979)145-79.

84 Trace Determination of N-Cyclohexyl-2-Benzothiazole Sulfenamide and 2-(4-Morpholinothio)benzothiazole in Water and Sediment by Gas Chromatography with a Flame Photometric Detector. S. Eto, R. Shinohara, A. Kido and T. Hori, Bunseki Kagaku, 29(3)(1980)213-16.

85 Direct Determination of Organic Compounds in Water Using Steam-Solid Chromatography. J. Teply and M. Dressler, J. Chromatogr., 191(1)(1980)221-9.

86 Extraction and Gas Chromatographic Determination of a Herbicide (Neburon) and its Metabolite(3,4-Dichloroaniline). Application to Natural Waters. A. Copin, J. Delmarcelle, R. Deleu, A. Renaud and Ph. Dreze, Anal. Chim. Acta, 116(1)(1980)145-52.

87 Trace Analysis of Organics in Water by Gas Chromatography. R. Mindrup, Pergamon Ser. Environ. Sci., 3(Anal.Tech.Environ.Chem.)(1980)325-33.

88 Gas-Chromatographic Analysis of the Soil Atmosphere. K.A. Smith, Adv. Chromatogr., 15(1977)197-231.

89 GC-MS in the Studies of the Aquatic Environment. R. Ishiwatari, Shitsuryo Bunseki, 26(4)(1978)301-11.

90 Determination of Very Low Concentrations of Primary Alcohols in Water. J. Tesar, V. Kubelka, J. Novak and J. Mostecky, Sb. Vys. Sk. Chem.-Technol. Praze (Oddil)F: Technol. Vody Prostredi, F23(1979)217-48.

91 Determination of Chromium(III) and Chromium(VI) by Gas Chromatography. S.R. Wang, F.Z. Xu, H.F. Zhou and X.L. Jin, Huan Ching K'o Hsueh, 1(3)(1980)11-16.

92 A Rapid Method for the Determination of Polyols in Cigarets and Leaf Tobacco. S. Ishiguro, T. Ohsumi, S. Matsushima and S. Sugawara, Kenkyu Hokoku-Nippon Senbai Kosha Chuo Kenkyusho, 121(1979)7-11.

93 Gas-Chromatographic Determination of Nicotine in Tobacco and Cigaret Smoke by Means of a Nitrogen-Phosphorus-Selective Detector. G. Lionetti and N. Carugno, Riv. Merceol., 19(2)(1980)99-111.

94 Utility of Mass Spectrometry in the Measurement of Trace Quantities of Hazardous Organic Species in Ambient Atmospheres. J.T. Bursey, M.D. Erickson, L.C. Michael, R.A. Zweidinger and E.D. Pellizzari, AIChE Symp. Ser. 76(196)(1980)338-45.

95 Determination of N-Nitrosamines from Diesel Engine Crankcase Emissions. E. U. Goff, J.R. Coombs, D.H. Fine and T.M. Baines, Anal. Chem.,52(12)(1980)1833-6.

96 Automated Sequential Sampler for Gas Chromatographic Determination of Trace Airborne Pesticides. H.L. Gearhart, R.L. Cook and R.W. Whitney, Anal. Chem., 52(13)(1980)2223-5.

97 Gas-Chromatographic Determination of the Components of Waste Water from Dimethyl Terephthalate Production. P. Kusz and A. Andrysiak, Fresenius' Z. Anal. Chem., 301(5)(1980)432-3.

98 Identification of Organic Compounds in a Mutagenic Extract of a Surface Drinking Water by a Computerized Gas Chromatography/Mass Spectrometry System (GC/MS/COM). W.E. Coleman, R.G. Melton, C. Frederick, K.A. Barone, T.A. Aurand and M.G. Jellison, Environ. Sci. Technol., 14(5)(1980)576-88.

99 Gas Chromatographic Analysis of Trace Hydrocarbon Pollutants in Water Samples. B.A. Colenutt and S. Thorburn, Int. J. Environ. Stud., 15(1)(1980) 25-32.

100 GC/MS Identification of Trace Organics in Philadelphia Drinking Waters During a 2-Year Period. I.H. Suffet, L. Brenner and P.R. Cairo, Water Res. 14(7)(1980)853-67.

101 Detection of Volatile Nitrosamines in Waste Water from Chemical Plants by Combined Capillary Gas Chromatography-Mass Spectrometry. G. Hartmetz and J. Slemrova, Bull. Environ. Contam. Toxicol., 25(1)(1980)106-12.

102 Gas Chromatographic-Mass Spectrometric Analysis of Chlorination Effects on Commercial Coal-Tar Leachate. K. Alben, Anal. Chem., 52(12)(1980)1825-8.

103 Highly Sensitive Detector for Amines and Their Derivatives. E.Y. Zandberg, A.G. Kamenev, V.I. Paleev, U. Kh. Rasulev, Zh. Anal. Khim., 35(6)(1980) 1188-94.

104 Volatile Phase Profiling of Mainstream Smoke by Glass Capillary Gas-Chromatographic Techniques. B.W. Good, M.E. Parrish and D.R. Douglas, HRC CC.J. High Resolut. Chromatogr. Chromatogr. Commun., 3(9)(1980)447-51.

105 Chromatographic Methods of Analysis for Methanol and Ethanol in Automotive Exhaust. C.J. Raible and F.W. Cox, SAE Tech. Pap. Ser. 790690(1979)7 pp.

106 Determination of Dimethyl and Diethyl Sulfate in Air by Gas Chromatography. J.C. Gilland, Jr. and A.P. Bright, Am. Ind. Hyg. Assoc. J., 41(6)(1980)459-61.

107 Construction and Testing of Bulk Filters with Coupled Gas Chromatographs for Determining Halogen Compounds in the Air. H. Hartkamp and A. Ionescu, Staub-Reinhalt. Luft, 40(4)(1980)151-6.

108 Determination of Particulate Organic Matter in Environmental Samples by Gas Chromatography-Mass Spectrometry and High Pressure Liquid Chromatography. R.C. Lao and R.S. Thomas, Pergamon Ser. Environ. Sci.,3(1980)97-110.

109 Quantitative Analysis of Methylchloride in the Atmosphere by Gas Chromatography with Electron Capture Detection. C. Vidal-Madjar, F. Benchah, M.F. Gonnord, J.P. Olivo and G. Guiochon, Pergamon Ser. Environ. Sci., 3 (1980)141-9.

110 Gas Chromatographic Methods for Atmospheric Sulfur Dioxide and Carbonyl Sulfide at the Parts-Per-Trillion Level. P.J. Maroulis, Diss. Abstr. Int. B, 41(3)(1980)946.

111 Characteristics of Determining Carbon Dioxide by Gas-Liquid Chromatography. I.D. Aziev, Akad. Nauk SSSR, Dagest. Fil., Inst. Geol., (Tr), 1(14)(1977) 101-3.

112 Coupling of Capillary Gas Chromatograph and Fourier Transform Mass Spectrometer. E.B. Ledford, R.L. White and S. Ghaderi. Anal. Chem., 52(14) (1980)2450-1.

113 A Recent Case of Environmental Pollution: Some Analytical Applications. A. Frigerio, Pergamon Ser. Environ. Sci., 3(1980)151-5.

114 Quantitative Determination of Naphthalenes in Tobacco Smoke by Gas Chromatography. R.F. Arrendale, R.F. Severson and M.E. Snook, Beitr. Tabakforsch. Int., 10(2)(1980)100-5.

115 Use of Thermoionic Detector in Gas Chromatography of Some Carbamate and Organophosphorus Insecticides. T. Shinohara and Y. Marumo, Kagaku Keisatsu Kenkyusho Hokoku, Hokagaku Hen, 33(3)(1980)136-9.

116 Gas-Chromatographic Method for the Determination of Traces of Carbon Disulfide in the Air. E. Saulova and M. Pankova, Khim. Ind. (Sofia), (4) (1980)164-5.

117 Application of a Direct Aqueous Acetylation Technique to the Gas Chromatographic Quantitation of Nitrophenols and 1-Naphthol in Environmental Water Samples. R.T. Coutts, E.E. Hargesheimer and F.M. Pasutto, J. Chromatogr., 195(1)(1980)105-12.

118 The Analysis of Organic Water Pollutants by Gas Chromatography and Gas Chromatography-Mass Spectrometry. R.A. Hites, Adv. Chromatogr., 15(1977) 69-112.

119 A Gas Chromatographic Determination of Benzotrichloride and Related Compounds in the Work Environment. H. Matsushita and S. Kanno, Ind. Health, 17(4)(1979)199-206.

120 Gas-Chromatographic Determination of Alcohols and Amide Solvents in the Biologically Purified Wastewaters from the Production of Fiber-Forming Polymer Materials. N.V. Makovskaya, B.I. Panchenko and I.N. Agranovskii, Khim Volokna, (3)(1980)53-4.

121 Determination of Polybrominated Biphenyl and Related Compounds by Gas-Liquid Chromatography with a Plasma Emission Detector. K.J. Mulligan, J.A. Caruso and F.L. Fricke, Analyst (London), 105(1256)(1980)1060-7.

122 Sample Preparation and Quantitative Determination of the Herbicide Atrazine (s-Triazine) from Water and Organic Matter: the Use of Gas Chromatographic Methods in Biological Experiments. G. Gunkel, Arch. Hydrobiol., Supp., 59(1)(1980)17-31.

123 Trace Analysis of N-Formylmorpholine in Hydrocarbons and Water by Gas-Liquid Chromatography. A.H. Tameesh, T.M. Sarkisian and K.R. Hameed, Chromatographia, 13(10)(1980)617-18.
124 Determination of Petroleum Gasoline and Benzene in the Air by Gas Chromatography. J. Krajewski and K. Nowicka, Med. Pr., 31(1)(1980)1-6.
125 Extraction-Free Gas-Chromatographic Determination of Small Concentrations of Organic Matter Dissolved in Water. V.Y. Botsan, V.I. Kofanov, Khim. Teckhnol. Vody, 1(1)(1979)59.
126 The Measurement of Airborne Benzene Vapor. H.G. Baxter, R. Blakemore, J.P. Moore, D.T. Coker and W.H. McCambley, Ann. Occup. Hyg., 23(2)(1980) 117-32.
127 A Convenient Method for the Determination of Ambient Nicotine. W.E. Crouse, L.F. Johnson and R.S. Marmor, Beitr. Tabakforsch. Int., 10(2)(1980)111-13.
128 Determination of Polycyclic Aromatic Hydrocarbons in Airborne Particulates at Various Sites in Taipei City by GC/MS and Glass Capillary GC. P.H. Chen, H.H. Shieh, J.M. Gaw, Proc. Natl. Sci. Counc., Repub. China,4(3)(1980)280-4.
129 Atmospheric Determination of Some Chlorinated Organic Products During Epichlorohydrin Production. D. Tolan, F. Pop, G. Kalmutchi and R. Popescu, Rev. Chim. (Bucharest), 31(6)(1980)596-7.
130 Modifications to Methods for Volatile Organics Analysis at Trace Levels. H.C. Hu and P.H. Weiner, J. Chromatogr. Sci., 18(7)(1980)333-42.
131 Gas-Chromatographic Analysis of Products of the High-Temperature Chlorination of Methane. T.S. Deineka and V.L. Senechko, Khim. Prom-st., Ser.: Metody Anal. Kontrolya Kach. Prod. Khim. Prom-sti., (9)(1980)7-9.
132 C_2-C_6 Hydrocarbons in Air: Gas Chromatographic Analysis of C_2-C_6 Hydrocarbons in Air Using a Concentrating Technique. E. Burghardt., Report 1978, IG-TNO-H-393; Sci. Tech. Aerosp. Rep., 18(14)(1980).
133 GC Analysis Aldehydes: Gas Chromatographic Analysis of C_2-C_6 Alkanes in Air Using an Isolation and Concentration Method Based on Bisulfate Adduction Formation. C. VanWaas, Report 1979, IG-TNO-H-399; Sci. Tech. Aerosp. Rep., 18(14)(1980).
134 Analysis of Waste Gases from Direct and Oxidative Chlorination. V.L. Senechko, Khim. Prom-st., Ser.: Khlornaya Prom-st., (1)(1980)13-14.
135 Selective Gas Chromatographic Determination of Chlorinated Hydrocarbons in Air with Preliminary Removal of Interfering Impurities. Y.S. Drugov, and G.V. Murav'eva, Zh. Anal. Khim., 35(7)(1980)1319-25.
136 Measurement of Phthalic Acid Esters in Air. H. Hayashi, T. Ishida, H. Suzuki, M. Nagata, S. Sato and M. Ichihashi, Kawasaki-shi Kogai Kenkyusho Nenpo, 6(1978)22-7.
137 Identification of Acephenanthrylene in Combustion Effluents. S. Krishman, R.A. Hites, Anal. Chem., 53(2)(1981)342-3.
138 Apparatus for Selective Injection of Sample Gases Into A Gas Chromatograph. Appl. 79/4,506, Jan.(1979) from Mitsubishi Monsanto Chemical Co., Gasukuro Kogyo, K.K. Jpn Kokai Tokkyo Koho 80 96,451.
139 Determination of Organochlorine Pesticides and Polychlorinated Biphenyls in Water by Gas Chromatography. J.D. Millar, R.E. Thomas and H.J. Schattenberg, III, Anal. Chem., 53(2)(1981)214-19.
140 Determination of the Concentrations of Organic Solvents Simultaneously Present in Air in the Workplace. M. Mukhtarova, Khig. Zdraveopaz, 23(2) (1980)173-7.
141 Distribution of Methylnaphthalene Isomers by Capillary GC-MS: a Means of Demonstrating a Component of Petroleum Origin in Complex Groundwater or Chemical Spill Samples. G.M. Frame, Control Hazard. Mater. Spills, Proc. Natl. Conf., (1980)180-4.
142 Assay and Recovery in Environmental Analysis. M.A. Bennett, Lab. Pract., 29(9)(1980)931.
143 Importance of Fractionation of Environmental Samples Prior to Analysis by High Resolution Gas Chromatography. K. Tjessem, Chem. Scr., 15(4-5)(1980) 189-92.

144 Determination of Organochlorine Insecticide Residues. II. Determination of α-, β-, γ-, δ-, ε-HCH and HCB in Water. S. Waliszewski, Fresenius' Z. Anal. Chem., 304(2-3)(1980)143.
145 A Method of Nitrogen Oxides Analysis by the Gas Chromatograph. S. Fukazawa, Y. Fujiwara and S. Tosaka, Hokkaido Kogyo Daigaku Kenkyo Kiyo, 6(1978)15-27.
146 Automated Gas Chromatographic Determination of Atmospheric Carbon Monoxide at the Parts-Per-Billion Level. A. Marenco and D.J.C. Delaunay, Anal. Chem. 53(3)(1981)567-70.
147 Estimated Application of Gas Chromatographic Headspace Analysis to Priority Pollutants. W.F. Cowen and R.K. Baynes, J. Environ. Sci. Health, Part A, A15(5)(1980)413-27.
148 Precise Determination of Dissolved Gases in Seawater by Shipboard Gas Chromatography. T. Gamo and Y. Horibe, Bull. Chem. Soc. Jpn, 53(10)(1980)2839-42.
149 Gas-Chromatographic Analysis of Trace Volatile Halogenated Hydrocarbons in Water (Drinking Water). Z.Y. Hu, M.C. Shih, T.F. Chen and F.C. Tao, Huan Ching K'o Hsueh, 1(5)(1980)38-43.
150 Enhancement of Electron Capture Detector Response to Polycyclic Aromatic and Related Hydrocarbons by Addition of Oxygen to Carrier Gas. D.A. Miller, K. Skogerboe and E.P. Grimsrud, Anal. Chem., 53(3)(1981)464-7.
151 Gas Chromatography of Simple Monocarbonyls in Cigaret Whole Smoke as the Benzyloxime Derivatives. D.F. Magin, J. Chromatogr., 202(2)(1980)255-61.
152 Development and Application of Methodology for Determining 1,2-Dibromo-3-Chloropropane (DBCP) in Ambient Air. J.B. Mann, J.J. Freal, H.F. ENos and J.X. Danauskas, J. Environ. Sci. Health, Part B, B15(5)(1980)519-28.
153 Determination of Atmospheric Organic Compounds by Capillary Gas-Phase Chromatography Coupled with Mass Spectrometry $(GC)^2(MS)$). M. Termonia, X. Monseur, G. Alaerts, A. DeMeyer, P. Dourte and J. Walravens, Comm. Eur. Communities (Rep) EUR, EUR(1980)6621.
154 National Emission Standards for Hazardous Air Pollutants; Alternative Test Method for Vinyl Chloride in Solvents and Poly(vinyl chloride) Resins. United States Environmental Protection Agency, Fed. Regist., 46(29)(1981) 12188-90.
155 An Improved Procedure for Sampling and Analysis of Dinitrotoluene Vapor Concentrations in Workplace Air. R.J. Hunt, N.R. Neubauer and R.F. Picone, Am. Ind. Hyg. Assoc. J., 41(8)(1980)592-4.
156 Gas Chromatographic Method for the Determination of Selenite and Total Selenium in Seawater. C.I. Measures and J.D. Burton, Anal. Chim. Acta, 120(1980)177-86.
157 Analysis of Trace Levels of Volatile Organic Contaminants in Municipal Drinking Water by Glass Capillary Gas Chromatography Using Simultaneous Flame Ionization and Electron Capture Detection. L.V. McCarthy, E.B. Overton, C.K. Raschke and J.L. Laseter, Anal. Lett., 13(A16)(1980)1417-29.
158 Determination of Organomercury Using a Combined Gas Chromatography-Mercury Analyzer. X. Zhang and X.G. Ren, Huan Ching K'o Hsueh, 1(6)(1980)34-7.
159 Determination of Nitriles in Environmental Samples. K. Numada, H. Tanaka and T. Kasai, Yamanashi-kenritsu Eisei Kogai Kenkyusho Nempo, 23(1980)65-70.
160 Gas-Chromatographic Determination of Substances Separating from Varnishes. T.D. Stepanenko, G.V. Frolova and L.M. Plesakova, Derevoobrab. Prom-st., (7)(1980)20-1.
161 Development of an Analytical Method for Fenitrothion and Five Derivatives in Water Using XAD Resins and Gas Liquid Chromatography (GLC). G.G. Volpe and V.N. Mallet, Int. J. Environ. Anal. Chem., 8(4)(1980)219-301.
162 Formation of Chloroform From the Chlorination of Diketones and Polyhydroxybenzenes in Dilute Aqueous Solution. S.D. Boyce and J.F. Hornig, Water Chlorination: Environ. Impact Health Eff., 3(1980)131-40.
163 Determination of Phthalic Anhydride in the Atmosphere of Workplaces by Gas Chromatography. A. Popler, Prac. Lek., 32(6-7)(1980)216-19.
164 Methods of Determining Micro Quantities of N,N-Dimethylaminopropionitrile in Air. V.E. Shefter, R.S. Krupenina and N.P. Ivanova, Zh. Anal. Khim., 35(9)(1980)1813-17.

165 Gas Chromatographic Determination of Ethyl Acrylate, 2-Ethylhexyl Acrylate and Methacrylic Acid in Air. S. Czerczak and T. Rogaczewska, Chem. Anal. (Warsaw), 25(3)(1980)459-64.
166 GC/MS Methodology for Measuring Priority Organics in Municipal Wastewater Treatment. D. F. Bishop, U.S. Environ. Prot. Agency, Off. Res. Dev., (Rep.) EPA 1980, EPA-600/82-80-196, 5 pp.
167 Room Temperature Chromatographic Determination of Helium-Argon-Nitrogen-Oxygen-Methane-Carbon Dioxide-Hydrogen Sulfide-Water in Underground Gas and Gas from Underground Water. H.F. Liu, C.L. Chang and T.H. Chang, Fen Hsi Hua Hsueh, 8(3)(1980)230-3.
168 Use of Two-Flame-Thermionic and Nitrogen Detectors for the Identification of Chemicals in the Air and Water. S.A. Volkov and E.E. Sotnikov, Gig. Sanit., (11)(1980)62-3.
169 Study of Hydrogen Sulfide Emissions from a Salt Water Marsh. A.B. Goldberg, P.J. Maroulis, L.A. Wilner and A.R. Bandy, Atmos. Environ., 15(1)(1981) 11-18.
170 Identification of Alcohols, Aldehydes, Ketones and Esters in Wastewaters. 2. Identification of Individual Organic Compounds. M.F. Prokop'eva and N.K. Tadzhieva, Deposited Doc. 1979, VINITI 3352, 16 pp.
171 Gas-Chromatographic/Mass-Spectrometric Analysis of Derivatized Amino Acids in Municipal Wastewater Products. J.L.Burleson, G.R. Peyton and W.H. Glaze, Environ. Sci. Technol., 14(11)(1980)1354-9.
172 Direct Analytical Method for Aliphatic Compounds in Water by Steam Carrier Gas Chromatography. K. Urano, K. Ogura and H. Wada, Water Res., 15(2)
173 Quantitative Determination of Dissolved Gases in Underground Gas or Water by Air-Standard Gas Chromatography. T.C. Chiang, L.C. Chao, K.C. Yuan, H.N. Yu and S.C. Li, Fen S Hua Hsueh, 8(4)(1980)351-3.
174 Gas-Chromatographic Determination of N,N-Dimethylformamide in Air Samples in the Micro Range. G. Boehm, W. Wliszczak and G. Kainz, Mikrochim. Acta 1(5-6)(1980)485-93.
175 Gas Chromatographic Method for the Determination of Dimethyl Sulfate in Air. K.S. Sidhu, J. Chromatogr., 206(2)(1981)381-3.
176 Determination of Halocarbons in Air by Gas Chromatography-High Resolution Mass Spectrometry. F. Bruner, G. Crescentini, F. Mangani, E. Brancaleoni, A. Cappiello and P. Ciccioli, Anal. Chem., 53(6)(1981)798-801.
177 The Determination of Some Chlorinated Organics in the Environment by Gas Chromatography-Mass Spectrometry. C.S. Creaser, Int. Environ. Saf., (Oct) (1980)29-31.
178 Analysis of Sludge Extracts by High Resolution GC with Selective Detectors. V. Lopez-Avila, HRC CC, J. High Resolut. Chromatogr., Chromatogr. Commun. 3(11)(1980)545-50.
179 Determination of Carboxylic Acids and Phenols in Water by Extractive Alkylation Using Pentafluorobenzylation, Glass Capillary GC and Electron Capture Detection. E. Fogelqvist, B. Josefsson and C. Roos, HRC CC, J. High Resolut. Chromatogr. Chromatogr. Commun., 3(11)(1980)568-74.
180 On-Column Injection in Gas Phase Chromatography-Capillary Columns. Application to the Analysis of Polycyclic Aromatic Hydrocarbons in the Environment. C.K. Huynh, Mit., Geb. Lebensmittelunters. Hyg., 71(4)(1980)FHJF.
181 Synthesis and Identification of the 10 Hexachlorodibenzo-p-Dioxin Isomers by High Performance Liquid and Packed Column Gas Chromatography. L.L. Lamparski and T.J. Nestrick, Chemosphere, 10(1)(-981)3-18.
182 Determination of Ligroine in Wastewater by Gas-Liquid Chromatography. T. Palaveeva and R. Daskalov, Maslo-Sapunena Prom-st., 16(1)(1980)32-5.
183 The Use of Porous Polymers for the Collection of Plant Volatiles. R.A. Cole, J. Sci. Food Agric., 31(12)(1980)1242-9.
184 Rapid Method for the Determination of Polycyclic Aromatic Hydrocarbons in Environmental Samples by Combined Liquid and Gas Chromatography. L. Szepesy, K. Lakszner, L. Ackermann, L. Podmaniczky and P. Literathy, J. Chromatogr., 206(3)(1981)611-16.

185 Determination of Nitrous Oxide in Flu Gas. T. Fujii, Kogai to Taisaku,16 (12)(1980)1187-91.

186 Some Critical Parameters in Collection, Recovery and Gas Chromatographic Analysis of Organic Pollutants in Ambient Air Using Light Adsorbents. G. Bertoni, F. Bruner, A. Liberti and C. Perrino, J. Chromatogr., 203(1981) 263-70.

187 Gas Chromatographic Determination of Mesityl Oxide and Diacetone Alcohol in the Air. M. Miazek-Kula, Pr. Cent. Inst. Ochr. Pr., 30(104)(1980)3-15.

188 Determination of Methylcyclopentadienylmanganesetricarbonyl by Gas Chromatography-Atomic Absorption Spectrometry at ng m^{-3} Levels in Air Samples. M. Coe, R. Cruz and J.C. Van Loon, Anal. Chim. Acta, 120(1980)171-6.

189 A Portable Photoionization GC for Direct Air Analysis. N.J. Barker and R. C. Leveson, Am. Lab. (Fairfield,Conn.), 12(12)(1980)76-8.

190 Gas Chromatographic Determination of Some Organic Compounds in Wastewaters. N.N. Shkilevich, L.K. Kanshaeva, M.T. Belova, Prom-st. Sint. Kauch.,10 (1980)9-10.

191 Gas-Chromatographic Determination of Acrylic and Methacrylic Acid Esters in Atmospheric Air. E.A. Komrakova and L.V. Kuznetsova, Gig. Sanit., (1) (1981)43-5.

192 Determination of Halogenated Hydrocarbons in Wastewaters by Head-space Gas Chromatography. C. Piccinini, Boll. Chim. Unione Ital. Lab. Prov., Parte Sci., 6(6)(1980)266-71.

193 PCBs in Transformer Fluids. N.A. Kirshen, VIA, Varian Instrum. Appl., 15 (2)(1981)10.

194 Analysis of Dissolved Gases. B. Thompson, VIA, Varian Instrum. Appl., 15 (2)(1981)11.

195 A Compact Automatic System for Measuring Carbon Dioxide and Ethylene Evolution by Harvested Horticultural Crops. A.E. Watada and D.R. Massie, HortScience, 16(1)(1981)39-41.

196 Trace Analysis of Fungicides in Grape Juice and Wine Using Glass Capillary Gas Chromatography. T. Spitzer and G. Nickless, HRC CC, J. High Resolut. Chromatogr. Chromatogr. Commun., 4(4)(1981)151-5.

197 Intercomparison of the Total Carbon Analysis, Flame Ionization Detection, and Gas Chromatographic Methods for Measuring Solvent Vapor Emissions. T.L. Riley, R.S. Marano, E. Chladek, J.H. Hoggett, S.P. Levine, J.M. Reinke, and R.W. Devlin, Proc., Annu. Meet.-Air Pollut. Control Assoc., 73rd(3) (1980) Paper 80-41.2, 16 pp.

198 Controlled Environment Portable Gas Chromatograph for In-situ Aircraft or Balloon-borne Applications. W.C. Kuster, P.D. Goldan and F.C. Fehsenfeld, J. Chromatogr., 205(2)(1981)271-9.

199 Determination of Lacquer Benzine C in the Air by Gas Chromatography. J. Krajewski and K. Nowicka, Med. Pr., 31(4)(1980)305-10.

200 Gas Chromatographic Detection of Sulfur Dioxide, Nitrogen Dioxide, Amines and Halocarbons Using an Aerosol Ionization Detector. P. Popp and G. Oppermann, J. Chromatogr., 207(1)(1981)131-7.

201 Application of Glass Capillary Gas Chromatography for Determination of Potential Hazardous Compounds in Workplace Environments. G. Becher, A. Bjoerseth and B. Olufsen, ACS Symp. Ser., 149(Chem. Hazards Workplace: Meas. Control)(1981)369-81.

202 Sampling and Determination of S,S,S-tributyl Phosphorotrithioate, Dibutyl Disulfide, and Butyl Mercaptan in Field Air. B.W. Hermann, J.N. Seiber, Anal. Chem., 53(7)(1981)1077-82.

203 Flame Ionization Detection. V. Danes, Czech. 184,565 (Cl. G01N23/00), 15 Aug.(1980), Appl. 76/2,782, 28 Apr(1976); 2 pp.

204 Hydrocarbon Analysis on a Receiving Stream. W. Kalbfus, Muench. Beitr. Abwasser-, Fisch.-Flussbiol., 32(Stoffhaushalt Fliessgewaessern),(1980) 187-94.

205 Gas-Chromatographic Determination of Nitrates and Nitrites in Water. A.L. Pertsovskii, T.V. Markovskaya and G.A. Kharnikova, Gig. Sanit.,(2)(1981)70.
206 Gas-Chromatographic Determination of Traces of Nitrate, Sulfate, and Phosphate in Aqueous Solutions. W. Faigle and D. Klockow, Fresenius' Z. Anal. Chem., 306(2-3)(1981)190-5.
207 The Use of an Inexpensive Electronic Filter for Improving Detection Limits in the Gas Chromatographic-Chemiluminescent Determination of Nitrosamines. M.E.B. Brown, D.L. James and J. Warren, Anal. Chim. Acta, 126(1981)125-33.
208 Chromatographic Determination of Water in Acetic Acid Products. G.S. Khzanyan, Zavod. Lab., 47(5)(1981)25-6.
209 Quantitative Determination of Volatile C_{2-6} Carboxylic Acids in Gases. A.I. Parimskii, U.S.S.R. 819,713 (Cl. G01N31/08), 07 Apr(1981), Appl. 2,655,787, 11 Sep(1978).
210 Identification of Polychlorinated Dibenzo-p-dioxins and Dibenzofurans in Environmental and Industrial Samples Using High-Resolution Gas Chromatography and Mass Spectrometry. H.R. Buser, Anal. Chem. Symp. Ser.(1979) (Pub.1980). 4(Recent Dev. Mass Spectrom. Biochem. Med., 6)515-21.
211 Current Status of Continuous Automated Measurement of Offensive Odor. 2. Continuous Measurement of Sulfur Compounds. T. Haseagawa, Seikatsu to Kankyo, 25(11)(1980)53-60.
212 Gas Chromatographic Determination of Nitrosamines in the Air. A. Audere, J. Dundurs, Z. Lindbergs, V. Berzins and A. Berzina, Gig. Tr. Prof. Zabol. (1)(1981)47-9.
213 Gas Chromatographic Measurement of N-Containing Compounds; Determination of Trimethylamine in Ambient Air with TENAX-GC Preconcentration and Chemiluminescent Nitrogen Detector-Gas Chromatography. N. Kashihira, K. Kirita, Y. Watanabe and K. Tanaka, Bunseki Kagaku, 29(12)(1980)853-8.
214 Identification of Odorous Compounds in Emissions from Municipal Sewage Plants by GC-MS. A. Zeman and H. Hagenguth, Proc., Annu. Meet. - Air Pollut. Control Assoc., 73rd(3), Paper 80-40.7(1980)16 pp.
215 Component Loss During Evaporation-Reconstitution of Organic Environmental Samples for Gas Chromatographic Analysis. W.D. Bowers, M.L. Parsons, R.E. Clement and F.W. Karasek, J. Chromatogr., 207(2)(1981)203-11.
216 A New Concept in Environmental Chromatography. G.F. Hewitt and J.N. Driscoll, Anal. Instrum., 19(1981)5-6.
217 Analysis of Volatile Organic Compounds in Water, Wastewater and an Industrial Effluent. P. Leinster, A.E. McIntyre, J.N. Lester and R. Perry, Chemosphere, 10(3)(1981)291-301.
218 Determination of Toxic Organic Components in Industrial Wastewater from the Volga-Don Industrial Complex. M.M. Evstifeev, A.I. Voloshina and Yu. M. Gavrilko, Izv. Sev. - Kavk. Nauchn Tsentra Vyssh. Shk., Tekh. Nauki, (4) (1980)21-2.
219 Gaseous and Volatile Liquid Hydrocarbons in the Marine Environment with Emphasis on the Gulf of Mexico. T.C. Sauer, Jr. and W. M. Sackett, Elsevier Oceanogr. Ser. (Amsterdam), 27A(Mar. Environ. Pollut., v1: Hydrocarbons)(1980)133-61, 531-68.
220 Separation of Aqueous Organic Mixtures by Pervaporation and Mass Spectrometry or Coupled Gas Chromatography-Mass Spectrometry. H. Eustache and G. Histi, Adv. Mass Spectrom., 8B(1980)1460-7.
221 Gas Chromatographic Analysis of Water Phenolic Pollutants Using Acid-Washed Graphitized Carbon Black. A. Di Corcia, R. Samperi, E. Sebastiani and C. Severini, Chromatographia, 14(2)(1981)86-8.
222 Determination of Polar Volatiles in Water by Volatile Organics Analysis. T. Ramstad and T.J. Nestrick, Water Res., 15(3)(1981)375-81.

CHAPTER 6

USE OF LIQUID CHROMATOGRAPHY IN ENVIRONMENTAL ANALYSIS

Due to its inherent lack of detection sensitivity for most compounds, liquid chromatography (LC) has not been the technique of choice in environmental analysis. The recent development of more sensitive detectors (e.g., the fluorometer), derivatization methods, and the necessity to utilize this technique for separations of complex mixtures, have aided rapid growth in environmental analysis methods utilizing LC.

This chapter outlines the approach to the analysis of environmental samples by liquid chromatography. While it cannot and will not catalog every environmental liquid chromatographic analysis, it will review major criteria for development of good separations, identification, and quantitation.

6.1 STANDARDS AND CALIBRATION

One of the most persistent problems in environmental analysis by liquid chromatography involves the availability of standards. Some commercial laboratories have made great advances toward solving this problem, but it seems that the most frequently requested analyses are for those compounds for which standards do not exist. Frequently, the material is available as a mixture (in a formulation of a herbicide, for example), or the material is a small component of a complex synthesis mixture. Standard materials can be obtained from traditional sources, such as the manufacturer, as well as from some supply houses. Table 2.1 (Chapter 2) identifies the sources for chromatographic standards.

6.2 SAMPLE INTRODUCTION ONTO THE COLUMN

The principle of sample introduction onto a liquid chromatographic column is the same as for a gas chromatographic column, namely, the sample should be introduced as a narrow plug to minimize peak broadening. For some liquid chromatographic columns, especially the small-particle columns, this principle is especially important since the high resolution inherent with the column can be negated by resolution losses from injection broadening forces.

6.2.1 Syringe injection

The least expensive (initially) and traditional method of sample introduction is with a syringe. Many syringe models have been designed for injection under high pressure, usually up to 5000 psi (35 MPa) without sample leakage or blow-by. Manual syringe injection, however, is the least reproducible means of sample introduction with a precision no better than two percent.

A very important component of a syringe injection system in liquid chromatography is the septum. Septa are usually fabricated from elastomeric, self-sealing materials, such as silicone rubber, neoprene, or fluoroelastomers. Some septa are made with a layer of Teflon to reduce attack on the material by the solvent. Even with these coatings, however, liquid chromatographic septa have shorter lifetimes than gas chromatographic septa since the conditions to which LC septa are exposed are more deleterious.

Some instruments have a septumless injector for liquid chromatography which avoids the problem of solvent attack. The system allows variable volume, high pressure injections, but the volume range and precision are limited somewhat. Since the injector plumbing remains in the flow path, the system is continously purged.

To avoid injection under high pressure, stopped flow injection has been used. In this method, injections are made onto the column under atmospheric pressure. The sample size is kept small to avoid band broadening, (< 5 μℓ) then the pump is turned on. The injection time is not critical, since liquid/liquid diffusion is not as great a problem as gas phase diffusion is in gas chromatography.

6.2.2 Valve injection

The most common mode of sample introduction in modern liquid chromatography is valve injection. Valve injectors are more convenient and more precise than syringe injectors and can be used for high pressure or high temperature injections. Long, narrow bore tubing is preferred for sample loops since narrow tubing decreases band width and thus does not degrade column efficiency (ref. 1). Reproducibility of larger sample volume (several milliters) up to 7000 psi (48 MPa) can give less than 0.2% error (ref. 2). Furthermore, automated analyses can be readily performed with automatic valve injectors. For environmental applications where many similar samples are involved, automated analyses are an important convenience.

6.3 SELECTION OF SEPARATION CONDITIONS

An advantage inherent in liquid chromatography which is not available in gas chromatography is the variety of both mobile and stationary phases for optimization of separation conditions. This degree of freedom adds more than its share to the difficulty in finding a good starting point. Table 6.1 gives a synopsis of the variety of special features for liquid chromatography. Figure 6.1 outlines the selection process for the LC separation's mode.

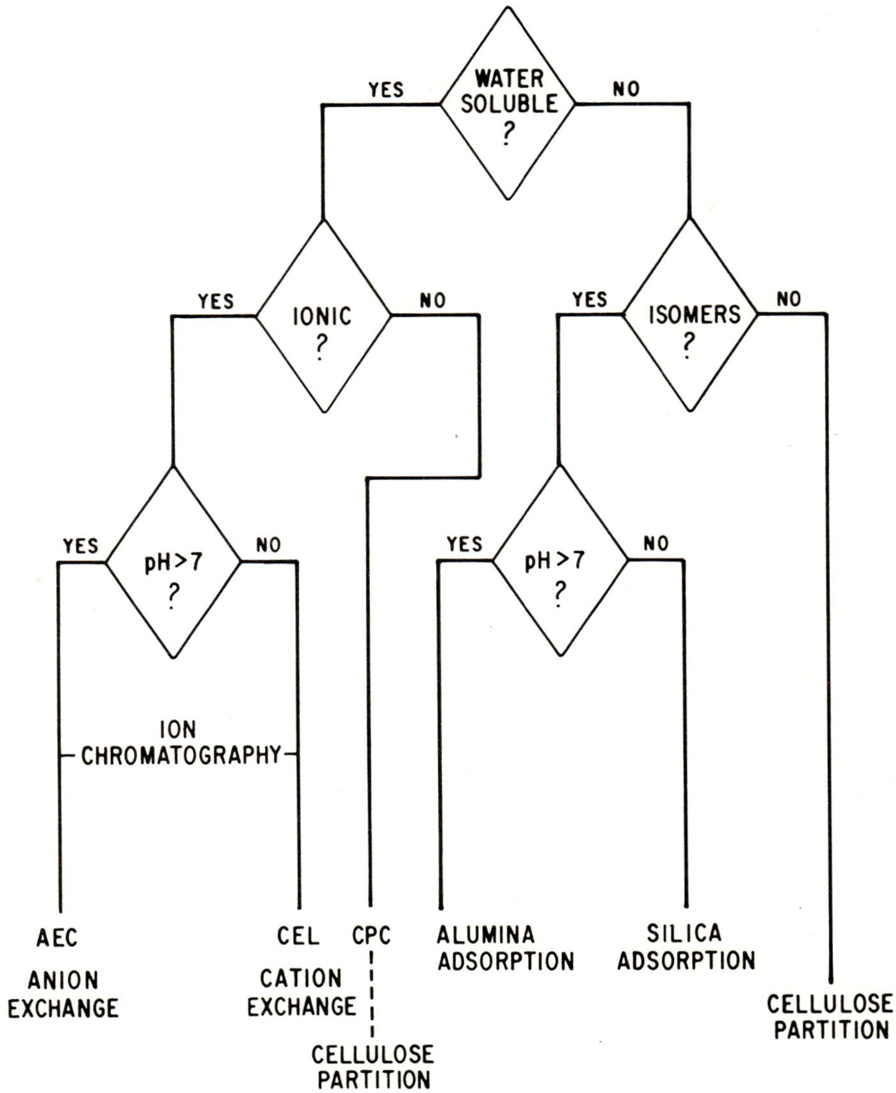

Fig. 6.1 Flow diagram for selection of mode of LC separation.

TABLE 6.1

Separation modes in liquid chromatography

Mode	Packing	Separation mechanism	Features
Liquid-Liquid LLC	Liquid coated on solid support	Partition	High selectivity (Caution-choose mobile phase that will not dissolve stationary phase).
Liquid-Solid LSC	Active solid	Adsorption	Good for isomer separations, stable. (Caution: can get irreversible adsorption).
Bonded Phase* BPC	Surface-reacted (chemically bonded) organic	Partition in combination with partition/adsorption.	BP not readily removed unlike LLC (Caution: wide variability in different manufacturer's products).
Ion Exchange IEC	Ion exchange resin or bonded phase particle	Ion exchange	Separated ionic species. Less efficient than other LC methods.
Ion Pair IPC	Liquid-coated support, solid, or bonded phase	Ion pair partitioning	Good for polar samples.
Size exclusion SEC	Organic gels or rigid particles	Exclusion of large molecules from pores.	Good for large molecule separations.
Sedimentation Field Flow Fractionation SFFF	None	Mobility of molecules in gravatational field	Good for large molecule separation.

*Note: Reversed phase chromatography uses a polar mobile phase and a less polar stationary phase, e.g., a C-18 bonded stationary phase and an acetonitrile/water mobile phase.

6.3.1 Water soluble samples

For samples containing a wide molecular size range size, size exclusion chromatography (SEC) is the method of choice. For molecules of similar size, bonded-phase chromatography (BPC) is the most useful and generally applicable.

For ionic species (ionizable), ion exchange or ion-pair chromatography may be tried. For samples in which ionization may be suppressed, the pH can be adjusted and the separation carried out as if the sample were not ionized. For strongly basic samples (pK_a above 6) the pH cannot be adjusted without causing degradation of most bonded phase packing material. Thus, ion-pair chromatography (IPC) must be used for this application.

6.3.2 Organic soluble samples

For samples containing a wide molecular size range, size exclusion chromatography (SEC) is the method of choice. For samples of low molecular weight (less than 2000) bonded phase or liquid solid chromatography is used.

For ionic (or ionizable) species where ionization may be suppressed by pH adjustment, bonded-phase chromatography can be applied. For non-ion suppressible samples, IPC or IEC is used.

For non-ionic species, LSC or BPC are used. The selection of the appropriate mode depends on whether isomers are present, LSC being the technique of choice for isomer separations.

6.3.3 Thin-layer chromatography (TLC) screening

Thin-layer chromatography is an often overlooked, relatively inexpensive means of separation. With the proper detection methods, a detection limit of one microgram is possible. Furthermore, multiple analyses can be completed in less than one hour.

Thin-layer chromatographic plates are available in the same chemical form as the conventional LC packings. Thus TLC can be used either as a screening process for environmental LC applications or for the analysis itself. Once the separation has been performed on a TLC plate, it is not difficult to transfer the information gained into a LC separation system (refs. 3 and 4).

6.3.4 Guard columns and precolumns

Two classes of contamination affect the life of a liquid chromatographic column: mechanical and chemical. Mechanical contamination causes blockage of the column and results from "dirty" samples or particles from the liquid chromatography system itself. These can be removed by proper sample "clean-up" by low pressure (column) liquid chromatography, filtration or centrifugation of the sample, or in-line filtration.

Chemical contamination results from irreversible reaction between components of the sample and the packing material. Often it is not possible to predict this problem and the column may be severely altered, necessitating its replacement. A solution to this costly problem is the installation of a guard column between the injector and the analytical column. Guard columns are packed usually with the same material as the analytical column and thus can trap any materials from the sample which might contaminate the analytical column. Guard columns sometimes are packed with another stationary phase in order to assist the analysis by selective adsorption or retardation of specific types or classes of compounds.

Guard columns can degrade chromatographic resolution either through inefficient packing or poor column connections. Guard columns are frequently dry-packed with pellicular, larger-diameter particles. The short columns can be emptied readily and repacked when necessary. Some chromatographers slurry pack guard columns, especially those containing porous packings, in order to increase overall resolving capabilities. These guard columns should be connected with minimum lengths of narrow, low-dead-volumne tubing. Narrow tubing causes less resolution loss than wide tubing.

Precolumns differ from guard columns in that these are usually located between the solvent pump and the injector. Precolumns do not degrade chromatographic resolution since they precede the injector. For slightly soluble liquid phases, precolumns are used to saturate the mobile phase with stationary phase liquid. For analyses where the pH and salt concentration are high, precolumns packed with large-particle, high surface area silica are used to saturate the mobile phase with silica. Both procedures are carried out to protect the analytical column.

6.3.5 Applications

The direct measurement of pollutants by liquid chromatography is not generally feasible due to the lack of sensitivity and the usually low concentrations of the materials of interest. Thus, concentration steps are usually employed using extractants or adsorbents; even an LC column itself (see Chapter 3) may be used. Frequently, additional sample treatment is performed for sample preservation, clean-up, and preliminary analyses (see Chapter 4). One of these steps may include a systematic separation by classes of the components of the sample, e.g., acid extractables, neutral extractables, etc. Every sampling location (from rural water supplies to industrial effluents) has its own percularities which makes methods development for pollutant analysis difficult. Thus, there can be no generally applicable scheme for all sampling locations.

The rest of this section contains specific information on pollutant analysis by liquid chromatography. Survey sources were Chemical Abstracts (ref. 5), NIOSH manuals (ref. 6) and Analytical Chemistry Reviews (ref. 7).

TABLE 6.2

Applications of liquid chromatography to environmental problems

Compounds(s), Type	Column(s)	Mobile Phase	Detector (λ in nm)	Comments	Ref.
PAH (polycyclic aromatic hydrocarbons)	C-18	0-100% Acetonitrile (ACN) in water	UV(202)	32 Priority pollutants	8
PAH	Al_2O_3	Cyclohexane; 97:3 cyclohexane: Et_2O	UV(230-470)		9

TABLE 6.2 (Cont.)

PAH	C-18	70:30 ACN in water	UV(254,280, 340,267,308)		10
PAH	C-18	20:80 water in ACN	UV(254); Fluorescence: λ_{ex} = 238; 265 λ_{em} = 370; 352; 420; 425		11
PAH in water	C-18	ACN in water	Fluorescence: λ_{ex} = 280; 305; λ_{em} = 340, 430, 500	16 PAH's	12
PAH	C-18	80:20 ACN in water	Fluorescence: λ_{ex} = 305; λ_{em} = 430	Selectivity factors	13
PAH in water	C-18	80:20 ACN in water	UV (254)	on graphitized carbon black	14
PAH in refinery water	C-18	70:30 methanol-water	UV (254)		15
PAH, phenolics, heterocyclics	C-18			Interlaboratory comparison	16
PAH	Polar bonded phases	-	UV	Comparison of 7 bonded phases	17
PAH	C-18	84-100% methanol in water	UV-scanning	TLC isolation	18
Nonionic surfactants	C-18	0-100% water in methanol	UV(280);M.S.		19

TABLE 6.2 (Cont.)

Phenol in air	C-18	85:15 methanol in water	UV(365)	p-nitro-benzeneazo-phenol derivative	20
HCHO in air	C-8	35:65 ACN in water	UV(340)	Converted to hydrazone	21
Aldehydes in air	C-18	30-42% ACN in water	UV(254)	Converted to hydrazone	22
Fatty acids in river water	C-18	40-100% ACN in water	UV(254)	Converted to phenacyl ester	23
Aromatic Amines	Cation exchange resin	10^{-3}-1M $HClO_4$ in water	UV(254)		24
NH_3 and alkyl amines in automobile exhaust		0.0025 NH_4NO_3	Conductivity	Ion chromotography (IC)	25
Isocyanates in working atmosphere	C-18	80:20(3) ACN in water (triethylamine phosphate) pH=3	UV(254); Fluorescence λ_{ex} = 254 λ_{em} = 412	Converted to urea derivative	26
Isocyanates in working atmosphere	C-18	75:25 ACN in water	UV(254)	Converted to urea derivative	27
Polychlorinated azobenzenes	C-18	80:20 ACN in water	UV(254)		28
4,4-methylene bis(2-chloroaniline)	C-18	80:20 ACN in water	UV(254)		29
NH_4^+ in ambient aerosols		0.006N HNO_3	Conductivity	IC	30

TABLE 6.2 (Cont.)

F^-, CrO_4^{-2}, NO_3^-, NO_2^-, NO_x, amines, azarenes			Conductivity	IC	31
SO_2 in atmosphere		0.003M $NaHCO_3$; 0.0024M Na_2CO_3	Conductivity	IC, 24-129/µg/m^3 SO_2	32
Cl, Br, P, S, I, F organic elemental		0.0030M $NaHCO_3$; 0.0024M Na_2CO_3	Conductivity	IC	33
Mg^{+2}, Ca^{+2}, Mn^{+2}, Fe^{+2}, Co^{+2}, Ni^{+2}, Cu^{+2}, Zn^{+2}, Cd^{+2}, Pb^{+2}, Bi^{+3}, Lanthanides	Strong acid ion exchange	Citrate, pH 4.6, 4.8	Vis(540)	Post-column reactor	34
Tl in natural waters	Anion exchange	SO_2 in water	Atomic absorption		35
Acrylic acid monomer	$-CN;-NH_2$	0.01% orthophosphoric acid in water	UV(195)		36
Fluoridone in water	C-18	60:40 methanol in water	UV(254)		37
Pesticides - a review					38
Pesticides - a review			116 pesticides		39
Carbamate pesticides in water	C-18	pH 6 aqueous phosphate buffer in ACN	Electrochemical		40
Nitrogenous organics	C-8	0.05M phosphate buffer (pH 6.9) or 0.05M borate buffer (pH 8.9) in methanol	UV(233); Fluorescence λ_{ex}=390; λ_{em}=470		41

TABLE 6.2 (Cont.)

32 priority pollutants	C-18	0-100% ACN in water	UV(202)	42
Priority pollutant phenols	C-18	30-80% ACN in water (1% acetic acid)	UV(280;254)	43
Carbaryl; 1-napthol	C-18	43:57 ACN in water	UV(222)	44
Pesticides	C-18	50:50 ACN in water	UV(214;229)	45
4,4'-Methylene bis (cyclohexyl isocyanate) in air	Silica	3:7:90 ethanol, isopropanol, isooctane	UV(230) aniline derivative	46

6.4 DETECTION OF SAMPLE COMPONENTS

The ideal liquid chromatographic detector for environmental substances is selective, has a wide linear range and high sensitivity, and is rugged and reproducible. Unfortunately, the ideal detector does not exist, but enough of these requirements are usually embodied in the more commonly used detectors, and these can be satisfactory for most specific applications.

There are two general categories of detectors, universal and selective. The universal detectors respond to changes in some bulk property of the effluent (e.g., changes in refractive index), whereas the selective detectors respond to some properties of the solute (e.g. absorption of a characteristic energy). For quantitation, the most important detector characteristics are stability, sensitivity, and the relationship of detector output to concentration. The latter factor must be frequently reconsidered, especially with selective detectors like the ultraviolet detector, where a minor constituent can give a larger detector signal than a major constituent in cases of large differences in molar absorptivity.

6.4.1 Refractive index detector

The refractive index (RI) detector was one of the first liquid chromatographic detectors. The RI detector responds to changes in the RI of the eluent, using the mobile phase as a reference. It is very sensitive to flow and temperature changes and suffers from a lack of sensitivity (10^{-6} g). In general, the RI detector is not used with solvent gradient, due to changes in the refractive index of the mobile phase. Although it is not truly a universal detector, the RI detector approximates a universal detector, since it responds to all compounds which have refractive indices different from the mobile phase. It is most frequently used in size exclusion chromatography for the detection of polymers and in preparative liquid chromatography.

6.4.2 Ultraviolet absorption detector

The ultraviolet absorption (UV) detector is the most commonly used detector in liquid chromatography. The UV detector responds to changes in the amount of light reaching a photometer. The quantity of light reaching the detector can be decreased by absorption of light by chromophores in the solute. The quantity of light absorbed is proportional to the concentration of solute and its molar absorptivity. The UV detector is very sensitive for highly absorbing compounds (10^{-11} g) and can be very selective, depending upon the wavelength monitored. Three types of UV detectors are available:

- single-wavelength (usually 254 nm)
- multiwavelength; and
- variable-wavelength.

The 254-nm band can be used for a wide variety of applications since many compounds absorb at or near that wavelength. The multiwavelength detector can be used to obtain qualitative information about a solute peak, since the absorption of various species varies with wavelength. The most versatile (and most expensive) UV detector is the variable wavelength detector which can be set to any wavelength in the UV-visible spectrum. The variable wavelength detector allows for the greatest flexibility, since it allows one to choose a wavelength for the best sensitivity and/or selectivity.

6.4.3 LC-GC detectors

Gas chromatographic detectors can also be used for some solutes in liquid chromatography. Gas chromatographs have been interfaced to liquid chromatographs to give a second dimension of separations power to the LC system. This capability can only be used for solutes that can be determined by gas chromatography alone, so that additional treatment of the LC peak may be needed to make it compatible with a GC system. GC detectors are much more sensitive, in general, than LC detectors (see Chapter 5), and can add further qualitative information.

A variation of the LC-GC technique of detection is the solute-transport-detector. This detector is comprised of a moving belt onto which the effluent is deposited. The volatile mobile phase is evaporated, and the nonvolatile solute is detected by some means such as a flame ionization detector. In this system, no additional solute separation is done and the mobile phase does not interfere with detection.

6.4.4 Infrared detectors

The infrared (IR) detector is similar to the UV detector in that it responds to changes in the light level due to the absorption of light by the solute molecules. IR detectors are not common due to lack of sensitivity (10^{-6}g at best) and the requirement that the mobile phase be free of any absorbing species. Since cells are usually constructed from salts, water or alcohols cannot be used in the mobile phase. Usually IR detectors are used in size-exclusion chromatography (SEC) and preparative liquid chromatography.

6.4.5 Fluorescence detectors

The fluorescence detector is one of the most sensitive and selective liquid-chromatographic detectors. This detector responds to light emitted by certain solutes when they are excited by high energy (UV) light. It is selective, since these special compounds will be excited only by certain wavelengths. The emitted light is measured by a photomultiplier at a 90° angle from the excitation source, so that the excitation light intensity is not measured. Thus, picogram (10^{-12}g) quantities of materials can be detected. Fluorescence detection has been used for the detection of polynuclear aromatic hydrocarbons and aflatoxins. In addition, fluorescent derivatives of nonfluorescing compounds may be made to enhance their detection limits.

6.4.6 Electrochemical detector

The most common form of electrochemical detector is the constant-potential-amperometric detector based on the mercury polarograph. As an electroreducible or an electrooxidizable substance is passed through the detector, the current changes. The magnitude of the current is proportional to the concentration of the electroactive species. Electrochemical detectors are most frequently used in reverse-phase chromatography (aqueous mobile phase), since the mobile phase must have high conductivity.

6.4.7 Thermal energy detector

The thermal energy detector has had great utility in the field of nitrosamine analysis. The N-NO bond in nitrosamines is thermally cleaved, and the NO radical produced reacts with ozone to produce excited nitrogen dioxide which emits light which is detected by a photomultiplier tube. This detector is very sensitive (10 ppt); and, although the thermal energy detector can react to other nitrogen-containing compounds, it is very selective.

6.4.8 Conductivity detector

The conductivity detector is used almost exclusively with ion-chromatography. It is based on the generation of an electrical current in the presence of ions. The second column used in most ion-chromatography removes the unwanted ions from the mobile phase and enables the solute ions to be detected in a low background solution. The conductivity detector is used with reverse-phase liquid chromatography. Table 6.3 summarizes ions detected by the conductivity detector.

TABLE 6.3

Ions detected by the conductivity detector in ion-chromatography

Inorganic ions	Inorganic ions	Organic ions	Organic ions
ammonium	magnesium	acetate	maleate
arsenate	monofluorophosphate	ascorbate	malonate
azide	nitrate	benzoate	methacrylate
barium	nitrite	butyrate	methyl amine
borate	o-phosphate	butylphosphate	methyl phosphonate
bromide	perchlorate	citrate	N-butyl amine
calcium	potassium	chloroacetate	oxalate
carbonate	rubidium	chloropropyl sulfonate	propionate
cesium	selanate		phthalate
chlorate	sodium	cyclohexyl amine	pyruvate
chloride	strontium	dibutylphosphate	sarcosinate
chromate	sulfate	dichloro acetate	succinate
thionate	sulfide	diethanol amine	tartrate
thiosulfate	tetrafluoroborate	dimethylamine	tetra-methyl-ammonium-bromide
fluoride	thiocyanate	ethyl amine	
hypochlorite	formaldehyde	tri-chloro acetate	
iodide		formate	tri-ethanol amine
lithium		gluconate	tri-ethyl amine
		glycolate	trimethyl amine
		hydroxy-citrate	tri-n-butyl amine
		lactate	

6.4.9 Derivatives

Since most common liquid chromatographic detectors are not very sensitive, derivatization is frequently used to enhance detectability. Derivation can be done before, during, or after injection onto the liquid chromatograph system. Table 6.4 summarizes common derivatives.

Fluorescence detection offers the greatest selectivity and sensitivity. Sensitivity rivals immunoassay techniques (see 6.4.5). Peak identification through fluorescence and emission spectra is a great potential advantage.

TABLE 6.4

Common derivatization reagents

Reagent	Class of compound (Reactant)	Comments
Dinitrobenzoyl-chloride(DNBC)	alcohols,amines,phenols	UV
Phenylthio-hybantoin(PTH)	amino acids	UV
p-nitrobenzyl-bromide(PNBB)	carboxylic acids	electrochemical, UV
p-nitrobenzyl-N-n-propylamine hydrochloride(PNBPA)	isocyanates	UV
N-succinimidyl p-nitro-phenyl-acetate (SNPA)	amines, amino-acids	UV
p-nitrobenzyloxyamine hydra-chloride (PNBA)	aldehydes, ketones	UV
o,p-nitrobenzyl-N,N'diisopropyl-isourea (PNDBI)	carboxyllic acids	UV
4-bromomethyl-2-methoxy-coumarin (BMC)	carboxylic acids	fluorescence
7-chloro-4-nitrobenzo-2-oxa-1,3 diazole (NBD - Chloride)	thiols, primary and secondary amines	flourescence
1-dimethylamino napthalene-5-sulfonyl chloride (Dansyl chloride	primary and secondary amines, phenols	fluorescence
1-dimethylamino napthalene-5-sulfonyl hydrazine (Dansyl hydrazine)	aldehydes	fluorescence
4-phenylspiro-[furan-2(BH),1'-phthalan]-3,3' dione (fluorescamine)	primary amines	fluorescence
Ce (IV)	reducing compounds	forms Ce(III)

REFERENCES

1. J. L. Glajch, D. C. Warren, M. A. Kaiser, and L. B. Rogers, Anal. Chem., 50 (1978) 1962.
2. L. R. Snyder and J. J. Kirkland, Introduction to Modern Liquid Chromatography, 2nd. ed., John Wiley & Sons, New York, 1979.
3. B. Cog, J. P. Nicolas, A. Lamotte, and M. Porthault, J. Chromatogr., 97 (1974) 137.
4. R. Amos and S. G. Perry, J. Chromatogr., 83 (1973) 245.
5. Chemical Abstracts, American Chemical Society, Washington, DC, January 1972 through October, 1981.
6. NIOSH Analytical Methods manuals, National Institute of Occupational Safety and

Health, Washington, DC, through volume 5.
7. Analytical Chemistry, bienniel reviews, Applications, 1979, 1981.
8. K. A. Pinkerton, JHRC & CC, 4 (1981) 33.
9. T. Dutkiewicz, J. Maslowski, S. Ryborz, N. Masny, Bromatol. Chem. Toksykol, 13 (1980) 89.
10. R. K. Sorrell and R. Reding, J. Chromatogr., 185 (1979) 655.
11. J. Chmielowiec and H. Sawatzky, J. Chromatogr., Sci., 17 (1979) 245.
12. K. Ogan, E. Katz, and W. Slavin, Anal. Chem., 51 (1979) 1315.
13. E. Katz and K. Ogan, J. Liquid Chromatogr., 3 (1980) 1151.
14. A. Lagana, B. M. Petronio, and M. Rotatori, J. Chromatogr., 198 (1980) 143.
15. H. G. Preston and A. Macaluso in Measurement of Organic Pollutants in Water and Wastewater, C. E. Van Hall, Ed., American Society for Testing and Materials, 686 (1979) 152.
16. H. S. Hertz, L. R. Hilpert, W. E. May, S. A. Wise, J. M. Brown, S. N. Chesler, and F. R. Guenther in Measurement of Organic Pollutants in Water and Wastewater, C. E. VanHall, Ed., American Society for Testing and Materials, 686 (1979) 291.
17. J. Chmielowiec and A. E. George, Anal. Chem., 52 (1980) 1154.
18. D. R. Choudhury and B. Bush, Anal. Chem., 53 (1981) 1351.
19. A. Otsuki and H. Shiraishi, Anal. Chem., 51 (1979) 2329.
20. K. Kuwata, M. Uebori, and Y. Yamazaki, Anal. Chem., 52 (1980) 857.
21. R. K. Beasley, C. E. Hoffmann, M. L. Rueppel, and J. W. Worley, Anal. Chem., 52 (1980) 1110.
22. K. Kuwata, M. Uebori, and Y. Yamasaki, J. Chrom. Sci., 17 (1979) 264.
23. D. A. Hullett and S. J. Eisenreich, Anal. Chem., 51 (1979) 1953.
24. J. O. Frohliger, K. S. Booth, N. Kotsko, American Chemical Society Symposium Series, 120 (1980).
25. R. B. Zweidinger, S. B. Tejada, J. E. Sigsby, Jr., and R. L. Bradow in Ion Chromatographic Analysis of Environmental Pollutants, E. Sawicki, J. D. Mulik, and E. Wittgenstein (Eds.), Ann Arbor Science Publishers, Inc., Ann Arbor, Michigan, USA, (1978) 125.
26. C. Sango and E. Zimerson, J. Liquid Chromatography, 3 (1980) 971.
27. C. Sango, J. Liquid Chromatography, 2 (1979) 763.
28. L. G. Richards, Altex Chromatogram, 2 (1978).
29. S. M. Rappaport and R. Morales, Anal. Chem., 51 (1979) 19.
30. J. D. Mulik, E. Estes, and E. Sawicki in Ion Chromatographic Analysis of Environmental Pollutants, E. Sawicki, J. D. Mulik, and E. Wittgenstein (Eds.), Ann Arbor Science Publishers, Inc., Ann Arbor, Michigan, USA (1978) 41.
31. E. Sawicki in Ion Chromatographic Analysis of Environmental Pollutants, J. O. Mulik and E. Sawicki (Eds:), 2 (1979) 1.
32. J. D. Mulik, G. Todd, E. Estes, R. Puckett, E. Sawicki, and D. Williams, in Ion Chromatographic Analysis of Environmental Pollutants, E. Sawicki, J. D. Mullik, and E. Wittgerstein (Eds.), Ann Arbor Science Publishers, Inc., Ann Arbor, Michigan, USA, 1 (1978) 23.
33. J. F. Colaruotolo in Ion Chromatographic Analysis of Environmental Pollutants, E. Sawicki, J. D. Mullick, and E. Wittgenstein, Ann Arbor Science Publishers, Inc., Ann Arbor, Michigan, USA, 1 (1978) 149.
34. R. M. Cassidy and S. Elchuk, J. Chrom. Sci., 18 (1980) 217.
35. J. Korkisch and I. Steffan, Int. J. Environ. Anal. Chem., 6 (1979) 111. Chem. Abstr. 91:128789k.
36. L. Brown, Analyst, 104 (1979) 1165.
37. S. D. West and S. J. Parka, J. Agric. Food Chem., 29 (1981) 223.
38. O. Matano and N. Shigy, Bunseki, 3 (1981) 157.
39. J. F. Lawrence and D. Turton, J. Chromatog., 159 (1978) 207.
40. J. L. Anderson and D. J. Chesney, Anal. Chem., 52 (1980) 2156.
41. N. M. Ram and J. C. Morris, J. Liquid Chromatogr., 4 (1981) 791.
42. K. A. Pinkerton, HRC & CC, 4 (1981) 33.
43. P. A. Realini, J. Chrom. Sci., 19 (1981) 124.
44. R. J. Bushway, J. Chromatog., 211 (1981) 135.
45. J. F. Lawrence, J. Chromatog., 211 (1981) 144.
46. R. A. Shoemaker, J. Chrom. Sci., 19 (1981) 321.

CHAPTER 7

SAFETY IN THE CHROMATOGRAPHY LABORATORY

The modern chemist has made immeasureable progress in recognizing and applying safe practices in the chemical laboratory. Today all sectors of the chemical world realize that immense personal and common benefits can be gained by greater emphasis on safe operation. An enlightened attitude is beginning to permeate every level of administration, creating a genuine interest in safety which is transmitted to all personnel through renovations of existing facilities and emphasis and construction of new facilities. Safety education programs and a vast wealth of literature are devoted to this subject. Some of the interest in safety has been reinforced by laws such as the United Kingdom Safety and Health Act and the United States TSCA (Toxic Substances Control Act), SDWA (Safe Drinking Water Acts), and OSHA (Occcupational Safety and Health Act), the German Federal Emmissions Control Act and related legislation. It is the aim of this chapter to point out specific safety considerations for the chromatography laboratory. By no means will this treatment be all-inclusive, but it is hoped that it will give fundamental information on which to build an extensive safety program.

Flammable liquids, exothermic reactions, unstable materials and toxic and corrosive materials play a large part in laboratory accidents and resulting injuries (ref. 1). A study of 100 significant laboratory fires by the National Fire Protection Association showed that 71% of the fires started within the laboratory, 56% started between 6:00 p.m. and 6:00 a.m., 21% were caused by electrical equipment and 20% by missuse of flammable liquids (ref. 2). The chromatography laboratory usually has more than its share of these potential hazards.

7.1 HAZARDOUS MATERIALS

7.1.1 Types

A hazardous materials may be defined as any substance which endangers health or life. Materials that may be found in the chromatography laboratory can be classified as flammable, corrosive (skin, eye, or respiratory irritant), toxic, or explosion hazard. Table 7.1 lists some common compounds usually encountered in the chromatography laboratory. The terms used in the table are described here.

Flash point: The temperature of the liquid at which sufficient vapor from the liquid is available to form an ignitable mixture. The flash point is generally considered the most significant parameter in assessing the flammable nature of liquids.

TABLE 7.1

Hazardous materials in the chromatography laboratory

[a]with permission, ACGIH, P.O. Box 1937, Cincinnati, OH, TWA;
[b]HCN liberated on contact with acid;
[c]suspected of inducing malignant neoplasms;
[d]fumes may decompose to phosgene when heated strongly;
[e]releases toxic cyanides when heated to decomposition.

Substance	Flash point(°C)	Boiling point(°C)	Ignition temp.(°C)	TLV(ppm)[a]	Comments
Acetone(2-propanone)	-17	56	538	1000	Hygroscopic
					Lachrymator[b]
Acetonitrile		82		40	
Benzene	-11	80	562	10	ACGIH (A2)[c]
Butanol-1	29	117	365	50	Irritant
Butylchloride		79			
Carbon disulfide	-30	46	100	20	Stench
Carbon tetrachloride		77		10	ACGIH(A2)[c,d]
Chloroform		61		25	ACGIH(A2)[c]
Cyclohexane	-20	81	260	300	-
o-Dichlorobenzene	74	180		50	-
N,N-Dimethylformamide		153		10	Irritant
Diethyl ether	-45	35	180	400	Peroxidizable
Dioxane	12	101	180	50	Peroxidizable[c]
Ethyl acetate	-4	77	427	400	-
Ethanol	13	78	393	1000	Hygroscopic
Heptane		98		400	-
Hexane	-22	69	261	100	-
Isopropanol		82		400	Hygroscopic
Methanol	11	65	464	200	Hygroscopic
Methylene chloride		40		200	-
Pentane		36		600	-
Pyridine		115		5	Irritant[e]
Tetrahydrofuran		66.0		200	Peroxidizable-irritant
Toluene	4	110.6	536	100	-
1,2,4-Trichlorobenzene		213.5			Irritant
2,2,4-Trimethylpentane		99.2			-

Boiling point: The temperature at which the vapor pressure of a liquid equals the atmospheric pressure. Ignition temperatures: The minimum temperature required to initiate or auto-sustain combustion in the absence of spark or flame.

TLV<(TWA): Threshold limit value (time weighted average): the time weighted average concentration for a normal 8-h work day or 40-h work week to which nearly all workers may be repeatedly exposed day after day without adverse effects.

Corrosive: A substance which acts to injure skin, eyes or surface tissues of the respiratory tract.

Toxic: A substance which may cause injury by direct chemical action with body tissues.

7.1.2 Use, disposal and storage of hazardous materials

At this time there are over 2,000,000 known compounds; the toxicity of most of these is not known. Over 6,000 compounds are considered by various groups to be cancer suspect agents or known carcinogens. Very few of these have been thoroughly tested. What can a chromatographer do when faced with an untested or unknown substance? Of course, the easiest answer is that, whenever possible, use an alternative material which is not hazardous. Of course, if an alternative is not possible, then it should be treated as if it were hazardous. The most important requirement when working with these materials is that as much information as possible be available to the analyst regarding physical properties, toxicology and safe practice. A general check-list of safe practices in the chromatography laboratory is given below:

Check list of safe practices:

(1) The laboratory should have a minimum of two exits.
(2) Labware should not be used for food or beverage.
(3) Smoking should not be allowed in the laboratory.
(4) Pipeting should be done mechanically, never by mouth.
(5) All bottles should be labeled properly.
(6) Only the minimum quantity of the material needed for the analysis should be present in the laboratory.
(7) Flammable solvents should be kept in safety cans, if possible.
(8) Trays large enough to hold the contents of the storage container should be under the container.
(9) Chemicals which may react with each other should be separated.
(10) Volatile or photo/heat-sensitive materials should be kept away from sunlight, switches, heating mantles, etc.
(11) Materials should be used in an approved hood.
(12) Personal protection should be worn (gloves, safety glasses, respirator, etc., depending upon the situation).
(13) Appropriate fire and first aid protection should be provided.
(14) All personnel should have frequent medical checkups.

TABLE 7.2

OSHA definitions for flammable and combustible liquids

Flammable		Combustible	
Class	Definition	Class	Definition
1A	Flash point below 22.8°C and boiling point below 37.8°C	11	Flash point at or above 37.8°C and below 60°C
1B	Flash point below 22.8°C and boiling point at or above 37.8°C	111A	Flash point at or above 60°C and below 93.3°C
1C	Flash point at or above 22.8°C and boiling point below 27.8°C	111B	Flash point at or above 93.3°C

(15) All personnel should be adequately trained with an ongoing safety program.

(16) Syringes should be stored with the needles secured.

(17) Compressed gases should be secured with a chain or cylinder stand. Flammable gases should not be stored with oxidizing gases.

Storage of flammable materials presents special problems especially to the small service laboratory. The liquid chromatography laboratory often needs large quantities of pure flammable solvents. These materials are best stored in fire-resistant flammable liquid storage vaults with automatic fire-extinguishing systems. If possible, a separate building should be used for bulk quantities of these materials. A ventilation system should be provided and all materials should be stored with spill-trays under breakable containers. (For good chromatographic practice, aromatics should be separated from other organic solvents.) Smoking should not be allowed near the cabinet. Toxic, explosive, corrosive or oxidizing agents should also be stored separately and have their own ventilating, spill, and fire-protection systems.

OSHA regulations (Occupational Safety and Health Act) and NFPA Standards (National Fire Protection Association) indicate that class 1A and 1B liquids may be stored in glass containers of volumes no larger than 1 gallon (3785 mℓ) if the purity would be affected in metal containers or if corrosion of the metal container would result (see Table 7.2 for definitions). These liquids should be stored in closed containers (Table 7.3).

If refrigerators are used to store solvents, special precautions should be taken to assure they are appropriate for the intended purpose. Household and commercial refrigerators, even if they are listed by Underwriters Laboratory (U.L.),

TABLE 7.3

Maximum size of storage containers

[a]tanks; [a]473.2 ml; [b]946.4 ml; [c]3785 ml.

Type	Class 1A	Class 1B	Class 1C	Class 11	Class 111
Glass	1 pint[a]	1 quart[b]	1 gallon[c]	1 gallon	1 gallon
Metal (other than Dot drums)	1 gallon	5 gallon	5 gallon	5 gallon	5 gallon
Safety cans	2 gallon	5 gallon	5 gallon	5 gallon	5 gallon
Metal Dot drums	60 gallon	60 gallon	60 gallon	60 gallon	60 gallon
Approved portable	660 gallon	660 gallon	660 gallon	660 gallon	660 gallon

were never designed for storage of flammables. "Safety" or "Explosion Safe" refrigerators should be certified by a professional agency. The refrigerator should be in compliance with current regulations.

Known carcinogens, on the other hand, require special conditions for use in storage (ref. 3). These should be used in regulated areas where only authorized persons can enter. No food, food containers, smoking, smoking materials or cosmetics should be allowed in the area. These areas should be at a lower atmospheric pressure than surrounding nonregulated areas. The materials leaving these areas must be checked for contamination. Safety equipment should be in or near the area where direct exposure could occur and personnel should be trained in their use. Medical surveillance of personnel may also be required.

Disposal of hazardous materials varies according to local laws, quantity of material, and hazard type. Burning, disposal through sewers, burying in chemical landfills and disposal by a commercial disposal company are common methods. Most liquid chromatography solvents can be incinerated and solute either burned or buried. The fate of these materials should be decided by a safety officer or a safety committee who should be aware of the legal and chemical aspects of the problem.

7.1.3 Compressed gases

Compressed gases pose a special hazard since they possess large quantities of potential energy - much greater than compressed liquid systems. Thus, compressed gases should be handled carefully as if they were potential rockets or fragmentation bombs. Some of the gases used in chromatography are flammable, corrosive, not life supportive, cold hazards, or toxic (Table 7.4). Cylinders filled with oxygen-poor substances and stored in confined spaces can cause anoxemia, making blood oxygen-deficient, resulting in unconsciousness and, in minutes, death (the ambient concentration of oxygen in breathing air need drop to only 15% to asphyxiate). Such confined spaces include cylinder closets, dry boxes, refrigerators, and cryogenic ovens. Liquid nitrogen and dry ice are the two most common cooling gases and may provide a cold hazard if bare skin comes in contact with these materials. Compressed gases have very high rates of effusion (hydrogen 2000 m/sec). Thus the laboratory can be permeated almost instantaneously if a leak occurs. Hydrogen which is used as a fuel gas for gas chromatography detectors and as a carrier for high-resolution gas chromatography could be an explosive hazard under some circumstances. In general few accidents have been attributed to these characteristics; however, one should be aware of their potential. Most accidents with cylinders result from physical strain in moving large cylinders. The check list (below) includes some hints on preventing cylinder mishaps.

TABLE 7.4

Properties of common compressed gases

[a] 1 MPa, 9.9 atm, 7.5×10^3 Torr;
[b] g-cal/(sec)(cm^2)(°C/cm) $\times 10^5$ at 0°C;
[c] United States Department of Transportation;
[d] ignition temperature 585°C;
[e] ignition temperature 540°C;
[f] use no grease or oil.

Gas	B.p(°C)	Crit.(°C) temp.	Crit:press (MPA[a])	Thermal[b] conductiv.	Physiological property	DOT[c] Label
Air	-189	-	-	5.6	Oxidant	Non-flamm.gas
Argon	-186	-122	4.9	3.9	Inert	Non-flamm.gas
Carbon dioxide	-78(sub.pt.)	31	7.4	3.4	Inert	Non-flamm.gas
Helium	-269	-268	0.23	33.6	Inert	Non-flamm.gas
Hydrogen	-253	-228	1.3	39.6	-	Flamm.gas[d]
Methane	-162	-82	4.6	7.2	-	Flamm.gas[e]
Nitrogen	-196	-147	3.4	5.68	Inert	Non-flamm.gas
Oxygen[f]	-183	-118	5.1	5.70	Oxidant	Non-flamm.gas

Cylinder safety check list

(1) Everyone in the laboratory should know if the contents of a cylinder are flammable, corrosive, life supportive, poisonous, or a cold hazard.
(2) The cylinder should be properly secured with chains or other stable support.
(3) The smallest size cylinder should be used for the job.
(4) Traps should be used to prevent back contamination of the cylinder.

(5) Proper regulators should be used with each cylinder. Regulator adapters should never be used.
(6) The main valve should be closed when the cylinder is not being used. The pressure in the regulator should be reduced to atmospheric pressure.
(7) Oxidizing and flammable cylinder should be separated, preferably by at least eight meters.
(8) Cylinders should not be stored near sources of heat.
(9) Steel cylinders should never be immersed in liquid nitrogen.

7.2 PERSONAL PROTECTION

Pliny the elder made the first reference to personal protection devices in the first century A.D. He recommended a crude bladder mask be worn over the faces of workers in cinnabar (HgS) operations to prevent inhalation of that toxic dust. Since then, slow, but steady, progress has been made, and a variety of devices is now commercially available. The quantity and type of these devices depends on the size and operation of the laboratory. In this section, emphasis will be given to those devices which may be useful in the chromatography laboratory.

7.2.1 Protective devices

Concern for the effectiveness of protective devices has increased rapidly in the last decade. Studies of "gas masks" and gloves for solvent handling have shown repeatedly the physical and chemical limitations of these devices. The American Society for Testing and Materials Committee F-23 continues to evaluate chemical protective clothing and suggest test methods for evaluation of chemical protective clothing.

The important consideration in selecting a protective device is the specifications for its intended use. One should also select a device whose specifications far exceed the worst possible case of intended use. Furthermore, one should not rely entirely on that device for protection since variations in its manufacture can change significantly its properties. For example, the permeability of protective gloves can change with changes in the thickness, filling agent, or plasticizer used.

Eye protection should always be worn in the laboratory. For most work, hardened glass or plastic lens provide adequate protection from mechanical hazards and chemical splashes. As the probability of splashes increases and when working with corrosive materials, side shields, safety goggles or full-face shields may be necessary. Under no circumstances should contact lens be allowed in the laboratory. These lens could intensify the injury from any splash since they prevent natural eye fluids from removing contaminants from the eye.

A new group has been organized to certify personal protective equipment made for the United States. The organization established is the Safety Equipment Institute (SEI), a non-profit organization established by the Industrial Safety Equipment Association, Arlington, Virginia (ref. 4). According to the association, the SEI program will be the first nongovernmental program to test and certify such equipment. The initial work involved certifying hard hats and protective eyewear. The organization was started because official regulatory channels were not meeting the needs for product certification. Only respirators, gas-detection tubes, sound-level meters and noise dosimeters now require certification by the National Institute for Occupational Safety and Health.

SEI will administer a three-part certification program for voluntarily submitted products, product-testing by independent laboratories, inspections of manufacturing facilities by an independent quality-assurance auditor and review of contested certification decisions by an impartial appeals board.

TABLE 7.5

Permeation of common solvents through glove materials after 0.5 h (ref. 5)

A: <0.1%; B: 0.1 - 1%; C: 1-10%; D: >10%; PVC = poly(vinyl chloride); PVA = poly(vinyl acetate) (ref. 5)

Solvent	Material					
	Natural rubber (0.4 mm)	Neoprene (0.4 mm)	Neoprene+ natural rubber (0.5 mm)	Nitrile (0.4 mm)	PVC (0.2 mm)	PVA (0.4 mm)
Carbon tetrachloride	D	D	D	A	D	A
Chloroform	D	D	D	D	D	A
Methylene chloride	D	D	D	D	D	A
1,1,2,2-Tetrachloroethane	D	D	D	C	D	A
1,1,2-Trichloroethane	D	D	D	D	D	A
Perchloroethylene	D	D	D	A	D	A
Methanol	A	A	A	A	B	D
Ethanol	A	A	A	A	B	C
2-Propanol	A	A	A	A	B	C
n-Butanol	A	A	B	A	B	A
Benzene	D	D	D	C	D	A
Toluene	D	D	D	C	D	A
Acetone	B	C	B	D	D	A
Methyl ethyl ketone	C	D	C	D	D	A
Tetrahydrofuran	D	D	D	D	D	A
Dimethylsulfoxide	A	A	A	A	B	D
Dimethylformamide	B	A	A	C	D	D
Pyridine	C	C	C	D	D	C
Dioxane	B	B	B	A	D	A
n-Hexane	C	A	C	A	D	A
Water	A	A	A	A	A	D

Hand protection is an area that is often ignored in the chromatography laboratory. Leather, wool-lined gloves or Kevlar® aramid gloves for handling hot-gas chromatographic apparatus should be located near each instrument. These gloves can also be used for mechanical protection in general handling of abrasives or glassware. Surgeons' gloves are useful for handling wet glassware. Corrosive materials should be handled with heavy rubber gloves to provide the best mechanical protection. Rubber gloves are often not very useful for solvent handling since they are permeable to many solvents. (According to NIOSH, protective clothing should resist permeation for at least 60 min (ref. 5)). Thus, a glove which is impermeable to the solvent should be worn over the rubber glove (which provides good mechanical protection) (Table 7.5). The outer glove should not be reused but placed in a special used-glove-disposal container.

Where heavy objects (such as gas cylinders) are transferred, foot protection is recommended. Usually steel-tipped safety shoes are sufficient for most chromatographic operations. Perforated shoes, sandals, or sneakers are not appropriate for laboratory work.

Laboratory coats or aprons should always be worn when working in the laboratory. These garments should be cleaned regularly and isolated from other garments.

Respiratory aids such as gas masks or self-contained breathing apparatus should be available. Self-contained devices are more reliable since canister gas masks are not universally effective. For non-emergency situations, airline masks that provide air from remote sources (e.g. bottled air, laboratory air supply) may also be useful. With these devices it is important that the particular filters be checked frequently. Many older chromatographic ovens were fabricated with asbestos. With these instruments it is important to keep dust levels as low as possible and to take special precautions when modifying the oven walls. The degree of protective equipment necessary will vary with the quantity and location of asbestos. Dust masks, respirators, and disposable head cover or coveralls may be considered.

7.2.2 Housekeeping

Good housekeeping is the key to safe laboratory operation. Not only does it allow for a more pleasant, organized work environment, but allows easy recognition of potential sources of danger. Good housekeeping includes not only the maintenance of a clean, orderly working area, but also the organization of procedures for processing samples, maintaining equipment and record keeping.

General housekeeping rules

(1) All bottles should be labeled before transfers are made or solutions prepared.
(2) Hoods should be storage places only for items that would be hazardous in the laboratory atmosphere.
(3) Hoods should be checked periodically with a velometer to be sure that they are effective for all hazardous materials (at least 100 ft/min).
(4) Syringes should be stored in drawers with needles covered and the drawers locked when not in use.
(5) Hands should be washed frequently, especially before eating or smoking.
(6) Gas cylinders should be clamped tightly in place.
(7) Long hair should be confined.
(8) All safety appliances (including visitors' safety glasses) should be available in convenient places.

(9) Emergency telephone numbers should be prominently displayed near the telephones.
(10) Housekeeping should be checked frequently by supervision or safety committee.

7.3 SAFETY LITERATURE

7.3.1 Books and pamphlets

"Matheson Gas Data Book", Matheson Gas Products, East Rutherford, NJ 1971.

"Carcinogens - Regulation and Control", US Dept. of Health, Education and Welfare, Public Health Service, Center for Disease Control, National Institute for Occupational Safety and Health, Division of Technical Services, Cincinnati, OH, 1977.

"Hazards in the Chemical Laboratory", Chemical Society, London, 1977.

"The Handbook of Reactive Chemical Hazards", by L. Bretherick, Butterworths, London, 1979.

"Patty's Industrial Hygiene and Toxicology", edited by G. D. Clayton and F. E. Clayton, Wiley, New York, 1979.

"The Merck Index", Merck & Co., Rahway, NJ, 1978.

"Dangerous Properties of Industrial Materials", edited by N. I. Sax, Van Nostrand Reinhold, New York, 1979.

"Registry of Toxic Effects of Chemical Substances", National Institute for Occupational Safety and Health and the American Chemical Society, Washington, DC, 1978.

"Documentation of the Threshold Limit Values for Substances in Workroom Air", American Conference of Governmental Industrial Hygienists, Cincinnati, OH, 1978.

"TLV's Threshold Limit Values for Chemical Substances and Physical Agents in the Workroom Environment with Intended Changes for 1979", American Conference of Governmental Industrial Hygienists, Cincinnati, OH, 1979.

"Guide for Safety in the Chemical Laboratory", Manufacturing Chemists Association, Van Nostrand Reinhold, New York, 1972.

"Fire Protection Handbook", National Fire Protection Association, Boston, MA, 1976.

7.3.2 Journals and papers

S. Altman, "Guidelines to Chemical Hazard Evaluation", J. Chem. Educ., 55 (1978) 140.

M. M. Renfrew, "Highlighting Safety Practices to Students", J. Chem. Educ., 55 (1978) 145.

D. O. Hedberg and E. Bussell, "Lab Safety Questionnaire", J. Chem. Educ., 55 (1978) 148.

F. W. Michelotti, "Hazardous Chemical Safety in the Laboratory", Anal. Chem., 51 (1979) 441A.

7.3.3 Films and audiovisual presentations

"28 Grams of Prevention" (film), Fisher Scientific Company, Pittsburgh, PA, 1978.

"Safety Attitudes" (slide), National Safety Council, Washington, DC.

"Using Fire Extinguishers the Right Way" (film), National Fire Protection Association, Boston, MA, 1978.

"Liquids Can Burn" (film), National Fire Protection Association, Boston, MA.

7.3.4 Miscellaneous

Toxic and Hazardous Chemicals in Industry Chart, Lab Safety Supply Co., Janesville, WI.

Chemical, Gas and Material Techniques and signs for the precautionary labeling of hazardous industrial chemicals, Lab Safety Supply Co., Janesville, WI.

Posters and Signs, School and Industrial Inspection Check list, National Fire Protection Association, Boston, MA.

Posters, National Safety Council, Chicago, IL.

REFERENCES

1 Case Histories in the Chemical Industry, Manufacturing Chemists Association, 1(1962), 2(1966), 3(1970).
2 Fire Record Bulletin FR88-3, National Fire Protection Association, Boston, MA.
3 Carcinogens - Regulation and Control DHEW Publication No. NIOSH77-206 U.S. Dept. of Health, Education & Welfare, CDC, NIOSH, Cincinnati, OH, 45226, Aug. 1977.
4 Chem. Eng. News, July 27 (1981) p. 10.
5 Chem. Eng. News, April 2 (1979) p. 15.
6 E. B. Sansone and Y. B. Tewari, "The Permeability of Laboratory Gloves to Selected Solvents", AIHA Journal, 39 (1978) 169.

CHAPTER 8

REGULATIONS, REGULATORY AND ADVISORY GROUPS

The creation of regulatory and advisory groups for issues of environmental concern has followed the development of legislation directed toward specific problems. These areas have included air, energy, animal and water resources, land use and waste disposal, transportation, noise, and historical preservation. It is not within the scope of this book to detail all the available information concerning these problems; however, this chapter will survey the major groups and regulations of international importance and point out only those regulation areas where either gas or liquid chromatography may provide significant measurement information.

8.1 GOVERNMENTAL AGENCIES, LEGISLATION, AND RELATED GROUPS

8.1.1 International organizations

8.1.1.1 UNEP (United Nations Environment Program) P. O. Box 30552, Nairobi, Kenya. This organization finances research and environmental programs. Funding comes from voluntary contributions and the UN general fund.

8.1.1.2 WHO (World Health Organization) Geneva 27, Switzerland. This organization was established in 1946 with the objective of attainment by all people of the highest possible level of health. It has a wide range of services including collection and dissemination of statistical data, establishment of standards and education.

8.1.1.3 CEC (Council of European Communities) - Program of Action on the Environment. The objective of this organization is constant improvement of the living and working conditions of the peoples of the European Economic Community.

On March 20, 1970, a directive on motor vehicle emissions was made by the Council. The directive stated that the mass of carbon monoxide and the mass of hydrocarbons must be less than the amounts given in Table 8.1. The directive also deals with characteristics of surface water intended for drinking purposes as given in Table 8.2.

8.1.2 National organizations

8.1.2.1 Canada. Department of the Environment: This agency has responsibility for protecting the country, environment and natural resources, although many provinces and municipalities have their own legislation. It is divided into the Atmospheric Environment Service, Environmental Protection Service, Fishery Services, Land, Forest, and Wildlife Services, Policy, Planning, and Research Service, Finance and Administration Service.

TABLE 8.1

Motor vehicle emission levels -CEC

Type 1 test - verifying the average emission in a congested urban area after a cold start.

Reference Weight (RW) (kg)	CO(g)	Hydrocarbons/test(g)
750	100	8.0
750 < RW < 850	109	8.4
850 < RW < 1020	117	8.7
1020 < RW < 1250	134	9.4
1250 < RW < 1470	152	10.1
1470 < RW < 1700	169	10.8
1700 < RW < 1930	186	11.4
1930 < RW < 2150	203	12.1
2150 < RW	220	12.8

TABLE 8.2

CEC Directive on Drinking Water Characteristics

G = Guide; 1 = mandatory.

A1 = simple physical treamtent and disinfection;
A2 = normal physical treatment, chemical treatment, and disinfection;
A3 = intensive physical and chemical treatment, disinfection, and extended treatment.

Parameter	A1		A2		A3	
	G	1	G	1	G	1
Nitrates (mg/ℓ NO_3^-)	25	50	-	50	-	50
Fluorides (mg/ℓ F^-)	0.7-1.0	1.5	0.7-1.7	-	0.7-1.7	-
Dissolved iron (mg/ℓ Fe)	0.1	0.3	1.0	2.0	1.0	-
Manganese (mg/ℓ Mn)	0.05	-	0.1	-	1.0	-
Copper (mg/ℓ Cu)	0.02	0.05	0.05	-	1.0	-
Zinc (mg/ℓ Zn)	0.5	3.0	1.0	5.0	1.0	5.0
Boron (mg/ℓ B)	1.0	-	1.0	-	1.0	-
Arsenic (mg/ℓ As)	0.01	0.05	0.05	0.05	0.1	0.05
Cadmium (mg/ℓ Cd)	0.001	0.005	0.001	0.005	0.001	0.005
Chromium (mg/ℓ Cr)	-	0.05	-	0.05	-	0.05
Lead (mg/ℓ Pb)	-	0.05	-	0.05	-	0.05
Selenium (mg/ℓ Se)	-	0.01	-	0.01	-	0.01
Mercury (mg/ℓ Hg)	0.0005	0.001	0.0005	0.001	0.0005	0.001
Barium (mg/ℓ Ba)	-	0.1		1.0	-	1.0
Cyanide (mg/ℓ CN^-)	-	0.05	-	0.05	-	0.05

TABLE 8.2 Cont.

Sulfates (mg/ℓ SO_4^{2-})	150	250	150	250	150	250
Chlorides (mg/ℓ Cl^-)	200	-	200	-	200	-
Phosphates (mg/ℓ P_2O_5)	0.4	-	0.7	-	0.7	-
Phenols (phenol index) (p-nitroaniline, 4-aminoantipyrene) (mg/ℓ C_6H_5OH)	-	0.001	0.001	0.005	0.01	0.1
Polycyclicaromatic hydrocarbons (mg/ℓ)	-	0.0002	-	0.0002	-	0.001
Dissolved/emulsified hydrocarbons (after extraction by light petroleum (mg/ℓ)	-	0.05	-	0.12	-	0.5
Total Pesticides (parathion, BHC, dieldrin) (mg/ℓ)	-	0.001	-	0.0025	-	0.005
Nitrogen (by Kjeldahl)	1.0	-	2.0	-	3.0	-
Ammonia (mg/ℓ $NH4^+$)	0.05	-	1.0	1.5	2.0	4.0

Examples of national legislation are given in Table 8.3. Specific limits for pollutants under Canadian laws are given in Tables 8.4 - 8.11.

TABLE 8.3

Canadian Legislation (National)

Water Act of 1970	Provided for joint Federal and provincial water basin planning, designation of water quality management areas, joint agencies for water quality management and commissions to conduct water resource management programs.
Federal Clean Air Act	Authorized the Federal Government to set air quality objectives and national emissions standards where there is a danger to health or where international agreements are involved.
Shipping Act	Prohibited pollution of the atmosphere by ships and protected against water pollution by discharges from vessels.
Motor Vehicle Safety Act of 1970	Defined characteristics of motor vehicle emissions.

TABLE 8.4

Exhaust emissions under the Motor Vehicle Safety Act (Canada)

Pollutant	Limit of pollution
Hydrocarbons	2.2 g/vehicle mile 275 ppm by volume for heavy-duty vehicle
CO	23 g/vehicle mile 1.5% by volume for heavy-duty vehicles

TABLE 8.5

Ambient air criteria under the Clean Environment Act of Manitoba

V = By volume, W = by weight.

Name of contaminant	Unit of concentration	Period of measurement	Concentration for commercial or land use	Concentration for residential or rural land use
SO_2	ppm in air (V)	1 h average	0.40	0.20
		24 h average	0.20	0.10
		annual average	0.05	0.02
Sulphation	mg SO_3/100 cm^2/ day dry (W)	30 days	1.0	0.4
Oxidants	ppm in air (V)	1 h average	0.15	0.15
		24 h average	0.10	0.10
		90% samplings of any one month	0.07	0.07
Oxides of Nitrogen	ppm in air (V)	1 h average	0.20	0.20
		24 h average	0.10	0.10
Beryllium	μg/^{3} in air (V)	24 h average	0.01	0.01
Fluorides	ppb in air as HF (V)	24 h average	4.0	1.0
Fluorides in forage	ppm in forage (W)	individual sample	-	35
Fluorides as gas	μg/100 cm^3/ 30 days (W)	30 days	-	40
CO	ppm in air (V)	1 h average	60	60
		8 h average	15	15

8.1.2.2 Austria. Federal Ministry of Health and Environmental Protection (Vienna): This agency has responsibility for federal environmental protection programs. According to the Austrian constitution, primary responsibility for environmental protection lies with the provincial and local governments.

8.1.2.3 Belgium. Ministry of Public Health and Environment (Brussels): This agency has responsibility for coordinating water, noise, and air pollution laws.

TABLE 8.6

Ambient water criteria under the Clean Environment Act of Manitoba

Contaminant	Concentration (mg/ℓ)[a]	Contaminant	Concentration (mg/ℓ)[a]
As	0.05	NO_3^- (in terms of N)	10
Cl	250-350	PO_4^{3-}	0.05
Cr(VI) (as Cr)	0.05	Se	0.01
Cu	1	Ag	0.05
CN^-	0.2	SO_4^{2-}	400
Fe	0.3	Zn	1
Pb	0.05	Mn	0.1
N (organic in terms of N)	1.5		

[a] Not to be exceeded

TABLE 8.7

Atmospheric emission criteria under the Clean Environment Act of Manitoba

Contaminant	Unit of concentration	Concentration point of impingement[a]
SO_2	ppm in air (V)[b]	0.3
CO	ppm in air (V)	5
Fluorides	ppb in air (V)	10
Be	$\mu g/m^3$ in air	0.01
Cl_2	ppm in air (V)	0.1
H_2S	ppm in air (V)	0.03
NH_3	ppm in air (V)	5
CH_2S	ppm in air (V)	0.15
HCl	ppm in air (V)	0.04
Pb	$\mu g/m^3$ in air	20

[a] Measured for 30 min.

[b] V = by volume

TABLE 8.8

Effluent limit criteria under the Clean Environment Act of Manitoba

Contaminant	Limit (mg/ℓ)	Contaminant	Limit (mg/ℓ)
As	0.5	Cl	the concentration which would increase the chloride content of the receiving body by 10
Cr	0.5	Cu	1
CN^-	0.25	Fe	to be considered in relation to the receiving body of water
Pb	0.5	Mn	10
N (Organic in terms of N)	20	NO_3^-	100
Oil	15	PO_4^{3-}	0.5
Se	0.1	Ag	0.5
Zn	1.0		

TABLE 8.9

Ontario standards for emitted contaminants ([a]Measured for 30 min.; [b]V = by volume)

Contaminant	Unit of concentration (in air)	Concentration at point of impingement[a]
NH_3	ppm (V)[b]	5.0
Be	$\mu g/m^3$	0.01
Br_2	ppm (V)	0.01
CdO	$\mu g/m^3$	10
CH_2S	ppm (V)	0.15
CO	ppm (V)	5.0
Cl_2	ppm (V)	0.1
Fluorides	ppm (V)	5.0
HCl	ppm (V)	0.04
HCN	ppm (V)	1.0
H_2S	ppm (V)	0.03
Fe	$\mu g/m^3$	10
Pb	$\mu g/m^3$	20
HNO_3	$\mu g/m^3$	65
Nitrogen oxides	ppm (V)	0.25
Ag	$\mu g/m^3$	1.0
SO_2	ppm (V)	0.3

TABLE 8.10

Ontario criteria for desirable air quality

V = by volume

Contaminant	Unit of concentration	Concentration at point of impingement	Period of measurement
Be	$\mu g/m^3$	0.01	24 hours
CO	ppm in air (V)	40	1 hours
		15	8 hours
		8	24 hours
Fluorides	ppb in air (V)	1.0	24 hours
		0.5	30 days
Fluorides in forage for compsumption by livestock	ppm in forage (W)	35 total	individual sample
Fluoridation(F^-)	$\mu g/100cm^2$	40 total	30 days
H_2S	ppm in air (V)	0.02	1 hour
Pb	$\mu g/m^3$	15	24 hours
		10	30 days
Oxidants	ppm in air (V)	0.10	1 hour
		0.03	24 hours
Oxides of nitrogen	ppm in air (V)	0.20	1 hour
		0.10	24 hours
Sulphation	mg $SO_3/100\ cm^2$	0.4/day	30 days
SO_2	ppm in air (V)	0.25	1 hour
		0.10	24 hours
		0.02	1 year

8.1.2.4 Denmark. Ministry of Environmental Protection (Copenhagen): This agency, created in 1971, consists of a secretariat, control and administration, planning and development and an Office of Environmental Protection with an expert and a service section. Local responsibility for pollution control lies with municipalities and local health courts.

8.1.2.5 Federal Republic of Germany. Ministry of the Interior (Bonn): This agency has the most important responsibility in the government for water management, air quality, and noise control. The VDI (German Engineering Society) proposes air quality limits, but has no legal power. The TAL (Technische Anleitung) is a Federal administrative regulation which must be followed (Table 8.12).

TABLE 8.11

Saskatchewan surface water quality criteria

Constituent	Maximum concentration (mg/ℓ)
Carbon Chloroform Extract (CCE) (includes Carbon Alcohol Extract: (CAE)	0.2

Constituent	Max. conc. (mg/ℓ)	Constituent	Max. conc. (mg/ℓ)
Methyl mercaptan	0.05	Na	30 - 75
Phenolics	0.005	S^{2-}	0.05
Resin acids	0.1	Zn	0.05
B	0.5	As	0.01
Cu	0.02	Ba	1.0
Chloride	1.5	Cd	0.01
Fe	0.3	Cr	0.05
Mn	0.05	CN^-	0.01
N (total organic	1.0	Pb	0.05
and inorganic)		Hg	0.0001
P (as PO_4^{3-})	0.15	Se	0.01
		Ag	0.05

8.1.2.6 France. Ministry for the protection of Nature and the Environment (Paris): This agency, organized in 1970, is responsible for chairing the interministerial Committee which administers regional and national parks, wildlife, hunting, and fishing activities and water pollution research. It also coordinates water pollution control programs and programs for protection of natural and historic sites. Pollution control is enforced by normal police action of the Prefecture. Laws have been initiated to enforce general policies and programs. Specific requirements are usually desired by the interministerial Committee subject to notification of the Ministry for the Protection of Nature and the Environment.

8.1.2.7 Iceland. Ministry of Health (Reykjavik): At present there is no centralized agency to coordinate environmental policy. The Ministries of Fisheries, Communication, Industries, and Social Affairs are responsible for pollution control efforts revelant to their domains.

8.1.2.8 Republic of Ireland. Ministry for Local Government (Dublin): This ministry is responsible for water pollution control policy, financing, planning, the control of local water and sewage programs, and the coordination of regional and local planning.

8.1.2.9 Italy. Ministry of Health (Rome): A subdivision of their agency, the Central Board of Control (which consists of representatives of several different

Departments), sees that the air pollution laws are enforced. This board was established in 1966. There are no water pollution laws.

8.1.2.10 Luxembourg. Ministry of the Interior, Public Health, and Public Works (Luxembourg Ville): This agency is responsible for nature conservation. The Police Department is responsible for enforcing the environmental law.

8.1.2.11 The Netherlands. Ministry of Public Health and Environmental Hygiene (Leidschendam): This agency is responsible for issuing licenses relating to surface water under state jurisdiction. Provincial administrators are responsible for other surface waters. The National Institute for Effluent Treatment is responsible for examining surface water quality and advising the Ministry on pollution prevention measures.

8.1.2.12 Norway. Ministry for the Protection of the Environment (Oslo): This ministry is responsible for protecting the air, soil, fresh and sea waters.

8.1.2.13 Sweden. National Environment Protection Board (Solna): This agency, which is responsible to the National Franchise Board and the Ministry of Agriculture, administers air, water, and noise pollution problems.

8.1.2.14 Switzerland. Office of Environmental Protection (Bern): This agency, part of the Department of Interior, is responsible for preparing legislation and overseeing its execution in the cantons, working with local authorities, advising the Federal government, informing the public, and taking part in national and international meetings and organizations.

TABLE 8.12

Gaseous emission limits in Germany (G.F.R.)

MIK_D = Highest permissible mean concentration with continuous exposure; MIK_K = highest permissible concentration with intermittant exposure (half hourly mean); TAL = maximum emission concentration of the Technische Anleitung; MAK = maximum allowable concentration at the work place.

[a] Three times per day.
[b] Once in 8 hours.
[c] Once in 2 hours.

Compound	MIK_D		MIK_K		TAL		MAK
	ppm	(mg/m^3)	ppm	(mg/m^3)	ppm	(mg/m^3)	(ppm)
Cl_2	0.1	(0.3)[a]	0.5	(1.5)[a]	0.3	(0.6)[b]	1
HCl	0.5	(0.7)[c]	1	(1.4)[c]	--	--	5
HNO_3	0.5	(1.3)[c]	1	(2.6)[c]	--	--	10
H_2S	0.1	(0.15)[a]	0.2	(0.3)[c]	0.15	(0.3)[b]	20
NO_2	0.5	(1)[a]	1	(2)[a]	1	(2)[b]	5
SO_2	0.2	(0.5)[c]	0.3	(0.75)[c]	0.4	(0.75)[c]	5
CO	--	--	--	--	--	--	50

Swiss Federal Institute of Technology (EAWAG): This research and teaching institute has no regulatory powers. It provides advisory services to the government and acts as an international reference center for waste management of WHO.

8.1.2.15 Japan. Environment Agency (Tokyo): This agency was established in 1971 to centralize and implement anti-pollution programs.

TABLE 8.13

Environmental quality standards for Japan

Compound	Daily average of hourly values[a]	Hourly values[a]
SO_2	0.04 ppm	0.1 ppm
CO	10.00 ppm	20.0 ppm[b]
NO_2	0.04 - 0.06 ppm	

[a] Not to be exceeded.

[b] In eight consecutive hours.

8.1.2.16 United States of America. The United States of America has the largest, most elaborate regulatory system for environment pollution analysis and control of any country in the world. There are hundreds of national, state, county, and municipal regulations pertaining to all aspects of environmental health and safety. In this section, major legislation and regulating bodies are outlined in order to give the reader an overview of these important laws and agencies.

A. Legislation

Occupational Safety and Health Act (OSHA) became effective April 28, 1971. Its purpose was "to assure safe and healthful working conditions for working men and women by authorizing enforcement of the standards developed under the act, by assisting and encouraging the states in their efforts to assure safe and healthful working conditions, by providing for research information, education and training in the field of occupational safety and health, and for other purposes." Under OSHA, the Federal government sets up mandatory safety and health standards. The DOL (Department of Labor) initiates and enforces the standards; the Department of

Health and Human Services implements educational and training programs.

TABLE 8.14

Maximum contaminant levels for inorganics from SDWA

Contaminant	Concentration (mg/ℓ)	Contaminant	Concentration (mg/ℓ)
Arsenic	0.05	Mercury	0.002
Barium	1	Nitrate (as N)	10
Cadmium	0.010	Selenium	0.01
Chromium	0.05	Silver	0.05
Lead	0.05		

TABLE 8.15

Maximum contaminant levels for organics from SDWA

Compound	Concentration (mg/ℓ)
Endrin (1,2,3,4,10,10-hexachloro-6,7-epoxy 1,4,4a,5,6,7,8,8a-octahydro-1,4, endo, endo-5,8-dimethuro napthalene)	0.0002
Lindane (1,2,3,4,5,6-hexachloro cyclohexane, gamma isomer)	0.004
Methoxychlor (1,1,1-trichloroethane); 2,2--bis[p-methoxyphenyl]	0.1
Toxaphene ($C_{10}H_{10}Cl_8$ - technical chlorinated camphene, 67-69% Cl)	0.005
2,4-D (2,4-dichlorophenoxyacetic acid)	0.1
2,4,5-TP Silvex (2,4,5-trichlorophenoxypropionic acid)	0.01

Mine Safety and Health Act (MSHA) became effective in 1977. It provided for the establishment of the Mine Safety and Health Administration in the DOL. Both OSHA and MSHA rely on NIOSH (National Institute for Occupational Safety and Health) for health and safety criteria.

Safe Drinking Water Act (SDWA) became effective in 1975. It sets requirements for contaminant levels, monitoring, and analysis of drinking water (Tables 8.14 and 8.15).

Clean Air Act (CAA) of 1963 with numerous amendments sets ambient air quality standards for a variety of pollutants. The EPA is charged with setting and checking these standards (Table 8.16).

TABLE 8.16

Primary and secondary ambient air quality standards from CAA

[a] Annual arithmetic mean.

[b] Maximum 24 h concentration not to be exceeded more than once per year.

[c] Maximum 3 h concentration not to be exceeded more than once per year.

[d] Annual geometric mean.

[e] Maximum 8 h concentration not to be exceeded more than once per year.

[f] Maximum 1 h concentration not to be exceeded more than once per year.

[g] Maximum number of days per calendar year < 1.

[h] Maximum arithemetic mean averaged over a calendar quarter.

Compound	Primary standard	Secondary standard
Sulfur oxides	80 $\mu g/m^3$ (0.03 ppm)[a]	1300 $\mu g/m^3$ (0.5 ppm)[c]
(SO_2)	365 $\mu g/m^3$ (0.14 ppm)[b]	
Particulate matter	75 $\mu g/m^3$[d]	60 $\mu g/m^3$[d]
	260 $\mu g/m^3$[b]	150 $\mu g/m^3$[b]
Carbon monoxide	10 $\mu g/m^3$ (9 ppm)[e]	40 $\mu g/m^3$ (35 ppm)[f]
(CO)		
Ozone	0.12 ppm (235 $\mu g/m^3$)[g]	0.12 ppm (235 $\mu g/m^3$)[g]
(O_3)		
Hydrocarbons	160 $\mu g/m^3$ (0.24 ppm)[c]	160 $\mu g/m^3$ (0.24 ppm)[c]
	(6 to 9 AM)	(6 to 9 AM)
Nitrogen dioxide	10 $\mu g/m^3$ (0.05 ppm)[a]	100 $\mu g/m^3$ (0.05 ppm)[a]
(NO_2)		
Lead	1.5 $\mu g/m^3$[h]	1.5 $\mu g/m^3$[h]
(Pb)		

B. Organizations

ACGIH (American Conference of Governmental Industrial Hygienists), P. O. Box 1937, Cincinnati, OH 45201 is an organization composed of only professional, industrial, governmental, or academic hygienists.

ACS (American Chemical Society) 1155 Sixteenth Street NW, Washington, DC 20036, publishes journals and sponsors a Division concerned with Environmental Chemistry.

AIHA (American Industrial Hygiene Association), 475 Wolt Leages Pky, Akron, OH 44311 is the accreditation organization for industrial hygienists.

AOAC (Association of Official Analytical Chemists), P. O. Box 540, Benjamin Franklin Station, Washington, DC 20044, USA is comprised of government and industry analytical scientists who test and sponsor improved methods for analyzing products related to health and agriculture.

APCA (Air Pollution Control Association), P. O. Box 2861, Pittsburgh, PA 15230, is comprised of individuals concerned with seeking economic answers to the problem of air pollution.

APHA (American Public Health Association), 1015 Fifteenth Street NW, Washington, DE 20036, is a professional association that seeks to protect and promote personal and environmental health.

ASTM (American Society for Testing and Materials), 1916 Race Street, Philadelphia, PA 19103, is a private society which develops voluntary consensus standards. It provides a legal, administrative and publications forum within which producers, users, ultimate consumers, and those representing a general interest can meet to write standards which best meet the needs of all concerned.

CMA (Chemical Manufacturers Association), 1825 Connecticut Avenue NW, Washington, DC 20009, is comprised of manufacturers of basic chemicals and is involved with public education, broad-based research, and promotes safety and emergency information programs.

ECE (Environmental Contaminant Evaluation) is a program sponsored by the United States Fish and Wildlife Service. Program participants perform research studies, monitoring, and field analyses to ascertain the effects of pesticides, industrial chemicals, heavy metals, and other pollutants on fish, wildlife, and their habitats.

EPA (Environmental Protection Agency), formerly the National Air Pollution Control Administration, is the agency charged with setting guidelines and establishing test procedures for the analysis of pollution.

MSHA (Mine Safety and Health Administration) is a government agency like OSHA, involved with the mining industry.

NIOSH (National Institute for Occupational Safety and Health) is a governmental agency which establishes criteria for health standards based on its research program. It is part of the Center for Disease Control, a unit of the U. S. Public Health Service, Department of Health and Social Services.

OSHA (Occupational Safety and Health Administration) is the branch of DOL responsible for implementing the Occupational Safety and Health Act.

8.1.2.17 U.S.S.R. The philosophy of Soviet legislation is that any atmosphere agent, even if it can not be proved as harmful, should not be permitted to threaten the individual. Thus air pollution limits are set below the level at which a physiological effect is felt.

TABLE 8.17

Soviet air pollution limits

Compound	Concentration $\mu g/m^3$	ppm	Time interval (h)
SO_2	0.05	0.02	24
	0.5	0.2	0.5
CO	1	0.9	24
	3	2.4	0.3
NO_2	0.085	0.045	24
HCl	0.2	0.15	0.3

8.1.2.18 Australia. Department of Environment and Conservation (Canberra): This Federal Department mediates legal problems. Air and water pollution control are the responsiblity of the state.

8.1.2.19 Israel Environmental Protection Service (Jerusalem): This agency coordinates environmental services, studies the problems, and makes recommendations on their solutions.

8.1.2.20 Mexico Environmental Protection Agency (Mexico, D. F.): This agency is part of the Ministry of Health. The ministry's power derives from the 1972 Laws to Prevent and Control Environmental Pollution, which designated this agency to study problems, implement methods, issue regulations, and start control measures.

8.1.2.21 New Zealand Commission for the Environment (Wellington): This agency administers water, soil, and air pollution problems.

8.1.2.22 South Africa: Three agencies located in Pretoria, are responsible for administering environmental laws: Department of Planning in the Environment, Department of Water Affairs, and the Department of Health.

8.2 LITERATURE

There are two types of literature with which the environmental chromatographer must be concerned - scientific and governmental. This section will give an overview of the most likely places one may find information useful to those concerned with air and water analyses by gas and liquid chromatography.

8.2.1 Scientific journals and periodicals

Advances in chromatography, J. C. Giddings, E. Grushka, J. Cazes, and P. R. Brown (Eds.), Marcel Dekker, New York, NY 10016, USA (from 1965)

American Industrial Hygiene Association Journal, AIHA, Akron, Ohio 44313, USA

American Water Works Association Journal, American Water Works Association, 6666 W. Quincy Avenue, Denver, CO 80235, USA

Analytical Chemistry, American Chemical Society, Washington, DC 20036, USA

Archives of Environmental Contamination and Toxicology, Springer, 175 Fifth Avenue, New York, NY 10010, USA

Archives of Environmental Health, Heldref Publications, 4000 Albemarle Street, NW, Washington, DC 20016, USA and Oxford OX3, OBW, England, Great Britain

Association of Official Analytical Chemists Journal, AOAC, P. O. Box 540, Benjamin Franklin Station, Washington, DC 20044, USA

Atmospheric Environment, Pergamon Press, Headington Hill Hall, Great Britain

Chromatographia, Vieweg, P. O. Box 5829, D-6200, Wiesbaden, G.F.R.

Critical Reviews in Environmental Control, CRC Press, 2255 Palm Beach Lakes Blvd., West Palm Beach, FL 33409, USA

Ecological Modeling, Elsevier Scientific Publishing Co., P. O. Box 330, Amsterdam, The Netherlands

Ecotoxicology and Environmental Safety, Academic Press, 111 Fifth Avenue, New York, NY 10003, USA

Environmental Health Prespectives, National Institute of Environmental Health Science, Department of Health and Human Services, U. S. Government Printing Office, Washington, DC 20402, USA

Environmental Pollution, Applied Science Publishers, Ripple Road, Barking, Essex, Great Britain

Environmental Research, Academic Press, 111 Fifth Avenue, New York, NY 10003, USA

Environmental Science and Technology, American Chemical Society, Washington, DC 20036 USA

European Journal of Toxicology, 49 Rue St. Andres-des Arts, Paris 75006, France

Industrial Water Engineerings Wakeman - Walworth, Inc., P. O. Box 1144, Darien, CT 06820, USA

International Journal of Environmental Studies, Gordon & Breach, 1 Park Avenue, New York, NY 10016, USA

Journal of the Air Pollution Control Association, APCA, 4400 Fifth Avenue, Pittsburgh, PA 15213, USA

Journal of Chromatography, Elsevier, Scientific Publishing Co., P. O. Box 330, Amsterdam, The Netherlands

Journal of Chromatography Science, Preston Publications, Niles, IL 60648, USA

Journal of Environmental Science and Health, Marcel Dekker, New York, NY 10016, USA

Journal of High Resolution Chromatography and Chromatography Communications, Hüthig, Heidelberg, G.F.R.

Journal of Liquid Chromatography, Marcel Dekker, New York, NY 10016, USA

Pesticides Monitoring Journal, EPA Office of Toxic Substances, U. S. Government Printing Office, Washington, DC 20402, USA

Pesticide Science Society of Chemical Industry Journal, Pesticide Science Society, London, Great Britain

Public Health Report, HEW Health Resources Administration, U. S. Government Printing Office, Washington, DC 20402, USA

Separation and Purification Methods, Marcel Dekker, Inc., New York, NY 10249, USA

Separation Science and Technology, Marcel Dekker, New York, NY 10249, USA

Toxicological and Environmental Chemistry Review, Gordon & Breach Science Publishers, 42 William IV St., London WC2, Great Britain

8.2.2 Scientific Books

Air Pollution Control, W. Strauss (Ed.), John Wiley, New York, 1971.

The Analysis of Air Pollutants, W. Leithe, Ann Arbor Science Publishers, Ann Arbor, MI, 1971.

Analytical Techniques in Environmental Chemistry, J. Albaiges (Ed.), Pergamon Press, Elmsford, NY, 1980.

Analytical Techniques in Occupational Health Chemistry, ACS Symposium Series No. 120. D. D. Dollberg and A. W. Verstufy (Eds.), American Chemical Society, Washington, DC, 1980.

Applied Headspace Gas Chromatography, B. Kolb, (Ed.), Heyden & Son, London, 1980.

Atmospheric Pollution 1980, M. M. Benarie (Ed.), Elsevier Scientific Publishing Co., Amsterdam, New York, 1980.

Chromatographic Analysis of the Environment, R. L. Grob (Ed.), Marcel Dekker, New York, 1975; Russian Translation, 1979.

Environmental Glossary, W. Frick (Ed.), Government institutes, Inc., Box 5918, New York, NY 10036, 1980.

Environment USA, R. R. Bowker Company, 1180 Avenue of the Americas, New York, NY 10036, 1980.

Instrumental Analysis for Water Pollution Control, by K. H. Mancy, Ann Arbor Science Publishers, Ann Arbor, MI, 1971.

Methods of Air Sampling and Analysis, American Public Health Association, Washington, DC 1972.

Paper and Thin-Layer Chromatographic Analysis of Environmental Toxicants, by M. E. Getz, Heyden & Son, London, 1980.

Priority Toxic Pollutants, M. Sitting (Ed.), Noyes Data Corp., Mill Rd. at Grand Ave., Park Ridge, NY 07656, 1980.

Water Pollution: A Guide to Information Sources, A. W. Knight and M. A. Simmond (Ed.), Gale Research Co., Book Tower, Detroit, MI 48226, 1980.

8.2.3 Regulatory information sources

Regulatory information sources can be obtained from the initiating body (e.g., the U. S. Federal Register). The organizations listed below give summary information on environmental regulations.

Environmental Research and Technology, Inc., Concord, MA 01742, USA, (bulletins)

Environmental Monitoring Series, Office of Research and Development, National Environmental Research Center, USEPA, Research Triangle Park, NC 27711, USA

NIH/EPA CIS (National Institute of Health/Environmental Protection Agency), Chemical Information System, Information Sciences Corporation, Washington, DC USA

Bureau of National Affairs, 1231 25th Street, NW, Washington, DC 20039, USA
- Chemical Regulation Reporter

Index to Government Regulation
- Chemical Substances Control
- Air Pollution Control
- Environment Reporter
- International Environment Reporter
- Occupational Safety and Health Reporter

Earth Law Journal, Academic Book Services Holland, Journal Dept., P. O. Box 66, Groningen, The Netherlands

Environmental Policy and Law, Elsevier Sequoia S. A., P. O. Box 851, 1001 Lausanne I, Switzerland

Environment Regulation Handbook, Environment Information Center, Inc., 292 Madison Avenue, New York, NY 10017, USA

8.2.4 Other Sources

Gas and Liquid Chromatography Abstracts, R. P. Taylor, (Ed.), Applied Science Publishers, Barking, Great Britain

Gas Chromatography Literature, Liquid Chromatography Literature, Preston Publications, Niles, IL 60648, USA

INFO terra, United Nations Environment Programme - a directory of potential sources of environmental information which can be accessed in over one hundred National Focal Points (NFPs)

Pollution Abstracts, Data Courier, Inc., 620 South Fifth Street, Louisville, KY 40202, USA

APPENDIX I

THE DETERMINATION OF ADSORPTION-DESORPTION EFFICIENCY OF CHROMATOGRAPHIC TRAPS

Chromatographic packing materials are used as adsorbents for air and water pollutants (see chapters 4-6). After sample collection, the pollutants are desorbed and then separated and quantitated by either gas or liquid chromatography. It is important that both the collection efficiency and the desorption efficiency of each component be known in order to obtain accurate and reproducible results. The outline below is given to help in the determination of these efficiencies.

The desorption efficiency of a compound is determined by injecting a known quantity of the compound onto the packed sampling tube. A minimum of three sampling tubes must be taken per concentration range studied. The quantity can be calculated directly and injected either neat or as a solution onto the tube. The tube is sealed and allowed to stand overnight to assure complete equilibrium. A second set of tubes should be set aside in a similar manner except that no sample should be added to any of the tubes (the blank). The compound is then desorbed (either by solvent or thermal desorption) and the blank tubes treated similarly. The blank, the desorbed compound, and a stock sample are analyzed in the same fashion as the unknown sample. The desorption efficiency is calculated from chromatographic data (peak areas) by the following equation:

$$\text{Desorption efficiency} = \frac{\text{area of desorbed sample-area of blank}}{\text{area of standard}} \times 100\%$$

The adsorption efficiency can be calculated by analyzing aliquots of the sample introduced onto the sampling tube. The peak areas obtained from the analysis of the aliquot added and the tube effluent give the adsorption efficiency. Each lot of sampling tubes should be tested. A second section of the tube or a back-up tube is frequently used to trap the effluent from the tested section.

Adsorption efficiency (constant load) $= \frac{\text{std. area} - (\text{effluent area} + \text{blank area})}{\text{standard area}} \times 100$

Calculations:

A. Quantity adsorbed in sample tube (Q)

$Q = 1000\ C_a V_s\ \rho$

Q	=	quantity adsorbed (µg)
C_a	=	concentration in air (ppm, µl/ℓ)
V_s	=	volume sampled(ℓ)(corrected to 25°C and 760 torr (0.1 MPa))
ρ	=	density (g/ml)

B. Concentration for calibration curved(C_F)

$C_F = \frac{FXV_sM}{24450pV_o}$

C_F	=	fractional concentration (µl solute/ml desorbing agent)
F	=	fractional quantities of air concentration (2,1,0.5...)
X	=	suspected air concentration (ppm, µl/ℓ)
V_s	=	volume samples (ℓ)
M	=	molecular weight (g/mole)
24450	=	molar volume (ml/mole) at 25°C, 0.1 MPa
ρ	=	density of solvent (g/ml)
V_D	=	volume of desorbing agent (ml)(e.g., CS_2)

Plot C_F versus peak height or peak area.

APPENDIX II

COMPARISON OF MODES OF SAMPLE INTRODUCTION IN CAPILLARY GAS CHROMATOGRAPHY

Since some methods of pollutants' analyses have required high resolution columns, a frequent question is what precision and accuracy can one obtain from capillary column studies. The most common problem in changing from packed to capillary columns comes from sample discrimination upon injection. There are four methods of introducing the sample onto a capillary column:

1. direct method,
2. split method,
3. splitless method,
4. cool on-column injection.

•Direct Method

The direct method uses a wide-bore open tubular column or a support-coated open tubular (SCOT) column. The direct method of injection allows one to put a larger sample onto the column; however, it has the disadvantage that it uses a lower resolution column than the traditional narrow-bore capillary column.

•Split Method

The split method of injection is the traditional mode of introduction onto the capillary column. A small portion (1:50, 1:100, 1:200, 1:300) of the sample goes onto the column. This method allows for high resolution separations; however, loss of quantitation can occur due to discrimination by boiling point in the splitting of the sample.

•Splitless Injection

This method takes advantage of the "solvent effect." Injection port temperature must be lower than that of the boiling point of the solvent. The solvent is condensed in the first few centimeters of the capillary column wherein the higher boiling impurities are concentrated. The temperature of the oven is raised and the splitter valve is turned on (usually after 0.5 minutes). This technique is not quantitative for compounds with a boiling point lower than that of the solvent.

• Cool-on-column injection

In the cool-on-column injection method, the sample is introduced without splitting in a liquid (not vapor) state directly onto the wide-bore column at a temperature near the boiling point of the solvent. Since the injection is made at a lower temperature, the possibility of decomposition and thermal rearrangements is minimized. This technique is the most accurate method of sample introduction in high resolution gas chromatography.

SUBJECT INDEX